Philosophical Perspectives on
Newtonian Science

Philosophical Perspectives on Newtonian Science

edited by Phillip Bricker and R. I. G. Hughes

A Bradford Book

Published under the auspices of
the Center for the History and Philosophy of Science of
The Johns Hopkins University

The MIT Press
Cambridge, Massachusetts
London, England

This book was set in Palatino by Achorn Graphic Services and printed and bound in the United States of America.

Library of Congress Cataloging-in-Publication Data

Philosophical perspectives on Newtonian science : papers presented to a conference at Yale University, November 6th and 7th, 1987 / edited by Phillip Bricker and R.I.G. Hughes.

   p.   cm.
  "Published under the auspices of the Center for the History and Philosophy of Science of the Johns Hopkins University."
   ISBN 0-262-02301-6
   1. Newton, Isaac, Sir, 1642–1727. Principia—Congresses.
2. Mechanics—Early works to 1800—Congresses. 3. Mechanics, Celestial—Early works to 1800—Congresses. 4. Science—Philosophy—Congresses. I. Bricker, Phillip. II. Hughes, R. I. G.
QA803.P55   1990
530'.01—dc20   89-29465
                                              CIP

Contents

Chapter 1
Philosophical Perspectives on Newtonian Science   1
R. I. G. Hughes

I

Chapter 2
On Locke, "the Great Huygenius, and the incomparable
Mr. Newton"   17
Howard Stein

Chapter 3
Foils for Newton: Comments on Howard Stein   49
Richard Arthur

II

Chapter 4
Real Quantities and Their Sensible Measures   57
Lawrence Sklar

Chapter 5
Absolute Time versus Absolute Motion: Comments on
Lawrence Sklar   77
Phillip Bricker

III

Chapter 6
Predicates of Pure Existence: Newton on God's Space and
Time   91
J. E. McGuire

Chapter 7
Newton on Space and Time: Comments on J. E. McGuire   109
John Carriero

IV

Chapter 8
Newton's Corpuscular Query and Experimental Philosophy   135
Peter Achinstein

Chapter 9
Reason and Experiment in Newton's *Opticks:* Comments on
Peter Achinstein   175
R. I. G. Hughes

V

Chapter 10
Kant and Newton: Why Gravity Is Essential to Matter   185
Michael Friedman

Chapter 11
The "Essential Properties" of Matter, Space, and Time: Comments
on Michael Friedman   203
Robert DiSalle

VI

Chapter 12
Ethical Implications of Newtonian Science   211
Errol Harris

Chapter 13
Modern Ethical Theory and Newtonian Science: Comments on
Errol Harris   227
Philip T. Grier

Contributors   241
Index   243

Preface

The papers collected here were delivered, more or less as they appear, at a conference held at Yale University on November 6 and 7, 1987, to honor the tercentenary of the publication of Isaac Newton's *Principia*. The conference was generously supported by the Ernst Cassirer Fund and the Commonwealth Foundation. We thank our co-organizer, John E. Smith, without whose efforts the conference would never have got off the ground. We thank the Yale Center for British Art for donating the use of their commodious facilities. We thank the Philosophy Department at Yale, and especially Pat Slatter, Assistant to the Chairman, for putting up with all our demands. Finally, we thank Peter Achinstein, who, on behalf of the Center for the History and Philosophy of Science at Johns Hopkins University, assisted in the publication of this book.

Phillip Bricker
R. I. G. Hughes

Chapter 1

Philosophical Perspectives on Newtonian Science

R. I. G. Hughes

i

Early in 1986 Halley's comet reappeared, as if to remind us that the tercentenary was at hand of the publication of one of the world's most remarkable and influential books, Sir Isaac Newton's *Philosophiae Naturalis Principia Mathematica*. It was a reminder in a double sense. Not only was Halley instrumental in the publication in 1687 of Newton's great work, to the extent of making himself financially responsible for its production, but the book itself contains a detailed analysis of the comet's motion.

This analysis appears in the third and last book of the *Principia*.[1] The preceding books deal, respectively, with the motion of bodies in a nonresisting medium, and with their motion in a resisting medium; the principles enunciated in these books are then used in Book III to "demonstrate the frame of the system of the world" (p. 397). Writing in 1942, the tercentenary of Newton's birth, E. N. da C. Andrade summarized the third book as follows:

> Not only does Newton establish the movements of the satellites of Jupiter, Saturn and the Earth, and of the planets round the Sun (or rather, as he points out, round the centre of gravity of the solar system) in terms of his gravitational theory, but he shows how to find the masses of the sun and planets in terms of the earth's mass, which he estimates quite closely; he accounts for the flattened shape of the earth and other planets; calculates the general variations of *g* over the surface of the earth; explains the precession of the equinoxes by consideration of the non-sphericity of the earth; calculates the main irregularities of the motion of the moon and of other satellites from the perturbing effects of the sun; explains the general features of the tides; and finally treats the orbits of comets in a way that shows that they are members of the solar system and enables the return of Halley's comet in 1759 to be accurately calculated.[2]

He goes on, "This brief and imperfect catalogue is only a reminder of the scope of this extraordinary book."

Andrade's "catalogue" is just part of a summary of the *Principia* contained in an address made to the members of the Royal Society. As he intended, the summary eloquently conveys the magnitude of Newton's achievement. But another and, in historical terms, equally striking testimonial is the fact that, more than two hundred years after Newton's death, Andrade could summarize and comment on the *Principia*, judiciously drawing attention to places here and there where Newton had gone astray, in a way that was immediately intelligible to an audience of practicing scientists.

His summary was intelligible for a simple reason: it dealt with the science that his audience had grown up with. Even now, three hundred years after the book's publication, the principles that Newton presented still form the framework within which basic physics is taught. The alert historian will wonder whether this identification of Newton's work and "Newtonian science" is not, perhaps, overly facile. I will postpone that question until later; either way, the answer will not affect the assessment that, to quote E. A. Burtt, Newton is "the one Englishman whose authority and influence in modern times has rivalled that of Aristotle over the late medieval epoch."[3]

Yet Burtt could also write of Newton that, "as a philosopher, he was uncritical, sketchy, inconsistent, even second rate."[4] How is the latter judgment to be squared with the fact of Newton's scientific preeminence? At the least, these hugely discrepant estimates prompt us to look again at Newton's work, at the philosophical issues it raises, and at the various ways in which Newton, his contemporaries, and his successors treated these issues.

The papers in this volume are responses to that invitation. They were given at a conference in the British Art Center at Yale University over a two-day period in the fall of 1987. Each has a particular focus, but the questions they deal with are all interlocked. My aim in this essay is to locate them within a philosophical picture of Newton's work, broadly sketched, and thereby to bring out some of the connections between them.

ii

To start with a truism: Newton's ideas were not formulated in an intellectual void. The synthesis of terrestrial and celestial mechanics that he effected marked the culmination of a scientific revolution which spanned nearly two centuries. This revolution did not consist merely in a transition from a geocentric to a heliocentric astronomy, and from an Aristotelian to a Newtonian dynamics, profound though those advances were; it also involved fundamental changes both in metaphysics and in scientific methodology. Nor, within this period, are we dealing with an incremental process in which individual contributors such as Kepler, Galileo, Descartes, and Newton all shared common assumptions about the practice of science. In particular, as we shall see, Newton's writings contain explicit repudiations not only of Descartes' theories but also of his methodology, and, even when he is not mentioned by name, Descartes can often be discerned lurking in the background of Newton's thought.[5]

Consider first the full title of the *Principia*—*Philosophiae Naturalis Principia Mathematica*—and compare it with Descartes' *Principia Philosophiae*.[6] Newton's title contains two qualifications absent from Descartes': the philosophy he treats is "natural philosophy," and the principles he educes are "mathematical principles."

In Newton's scientific writings there is virtually no "first philosophy," as that term is used by Descartes (*Principles*, p. xxv). The Preface to the *Principia* includes just three lines in which Newton talks explicitly about epistemological matters: "For the whole burden of philosophy seems to consist in this—from the phenomena of motions to investigate the forces of nature, and then from these forces to demonstrate the other phenomena" (pp. xvii–xvciii), and the text proper begins with a definition of "quantity of matter" (p. 1). The opening of the *Opticks* is equally laconic: "My design in this book is not to explain the properties of light by hypotheses, but to propose and prove them by reason and experiments: in order to which I shall present the following definitions and axioms."[7] Where Newton does present epistemological principles, as in the "Rules of Reasoning in Philosophy" at the start of Book III of the *Principia* (pp. 398–400), they refer exclusively to natural or "experimental" philosophy.

True, natural philosophy can lead to knowledge of God—indeed, as we shall see, God is all-pervasive in Newton's universe—but his thought moves in the opposite direction to that of Descartes. In Book I of Descartes' *Principles* God is mentioned twice in §5 (p. 4), and his existence is proved by the ontological argument in §14 (p. 8) and by the causal argument in §20 (p. 20). More important for the point I want to make, however, is the way that God appears in Book II, "Of the Principles of Material Objects." He is our guarantee that material objects exist (§1, p. 40), and in §33 Descartes argues "that God is the primary cause of motion: and that He always maintains an equal quantity of it in the universe" (p. 130). In contrast, neither in the first edition of the *Principia* nor in the first edition of the *Opticks* is there any mention of God.[8] Then in 1713, with the addition of the "General

Scholium" to the second edition of the *Principia,* the existence of God is adduced by the argument from design: "This most beautiful system of the sun, planets and comets, could only proceed from the counsel and dominion of an intelligent and powerful Being" (p. 544). The next two pages are then taken up with theological discussion. Similarly, the arguments concerning God that appear in the second edition of the *Opticks,* published in 1717, are arguments *to* the existence of God *from* the principles of natural philosophy. In this edition the first time the name God is used is on p. 400, six pages before the book ends; but thirty pages earlier Newton has written: "Does it not appear from phenomena that there is a Being incorporeal, living, intelligent, omnipresent, Who in infinite space, as it were in His sensory, sees the things themselves intimately, and comprehends them wholly by their immediate presence to Himself."[9]

On the relation between God and space, more later; my immediate concern is to point out that, whether or not Newton's science is in fact epistemically prior to his theological convictions, natural philosophy is, in McGuire's phrase, *functionally autonomous* both in the *Principia* and in the *Opticks.* Further, it is autonomous with respect not just to theology but to all "first philosophy."

iii

This refusal to lock natural philosophy into a preexisting system, whether rationalist or narrowly empiricist, Howard Stein presents as Newton's prime philosophical virtue. As the title suggests, his paper, "On Locke, 'the Great Huygenius and the incomparable Mr. Newton,' " compares Newton with two of his near contemporaries, Christian Huygens (1629–1695) and John Locke (1632–1704). Huygens' work is acknowledged by Newton (pp. 22, 25), and Stein looks at his derivations of the laws governing collisions between bodies and of the force required to constrain a body to move in a circular path.

Of particular interest here are Huygens' truly original methods of proof. From our perspective he appears as the first physicist to give a precise and accurate formulation of the Galilean principle of relativity. That principle asserts that all frames of reference moving with uniform velocity are dynamically equivalent, and Huygens makes unprecedentedly subtle and powerful use of it in solving dynamical problems. But Stein also suggests that Huygens' work was inhibited, as Newton's was not, by too rigid an adherence to the philosophical program of "mechanical philosophy," to the demand, that is, that all physical changes must be explained in terms of collision processes.

Similarly, in the second half of his paper Stein examines Locke's comparative dogmatism regarding the sources of our knowledge and the question of primary qualities and compares it with Newton's flexibility, to the latter's advantage.

Richard Arthur's "Foils for Newton" can profitably be read as a prolegomenon to Stein's paper. Arthur voices one reservation: that perhaps Stein overestimates the extent to which Newton would regard his "principles" as revisable. He points out, however, that even if this reservation is granted, Stein still provides a detailed and effective rebuttal of Burtt's view of Newton's philosophical competence.

iv

Recall that Newton's title contains a second qualification: the principles of natural philosophy are to be *mathematical*. Newtonian science is mathematical in (at least) two linked ways. In the first place, it is laid out in the Euclidean manner: propositions are deduced from definitions and axioms in accordance with mathematical canons of proof. Second, and the reason why this mode of presentation is so successful, it investigates aspects of nature that admit a mathematical representation, and assumes them to be fundamental. Thus the *Principia* deals exclusively with physical *quantities* (mass, force, velocity, and so on), and the *Opticks* offers a geometrical analysis of the nature of light. Both, one might say, endorse Galileo's dictum that the Book of Nature is written in the language of mathematics.[10]

Among the perceptible qualities of objects, Galileo tells us, only the mathematical qualities—number, figure, magnitude, position, and motion—correspond to qualities inhering in the things themselves. For Galileo these are the true *primary qualities. Secondary qualities,* "tastes, odors, colors and so on, are no more than mere names so far as the object in which we place them is concerned, and . . . reside only in the consciousness."[11] Only the primary qualities are within the province of science because only these qualities inhere in the objective world.

As we shall see, in the *Principia* Newton extends Galileo's list of primary qualities.[12] Nonetheless, any science that includes those qualities requires a framework of space and time with respect to which they can be specified. But how is this framework to be set up? Galileo used his pulse to time the descent of a ball down a ramp, but the measure of time that this provided could have been, at best, only approximately uniform. Similarly, as Newton points out (pp. 7–8), even astronomical measures of time have to be corrected by the "equation of time." And, with regard to space, since the motions we observe are all relative motions, on what grounds is one frame of reference to be preferred to any other?

Newton's answers to these problems have aroused as much controversy as anything else in the *Principia*. In the Scholium to the Definitions he famously—or notoriously—posits an absolute time with respect to which our measures of time are to be corrected, and an absolute space against which the "true and absolute motion of [a] body" takes place (p. 7).

A strong objection to the latter is that absolute space appears to play no role in Newtonian dynamics. As we have noted, the theory obeys a Galilean principle of relativity: all frames of reference moving with uniform linear velocity are dynamically equivalent. Thus Leibniz writes: "The fiction of a material finite universe, moving forward in an infinite empty space cannot be admitted . . . ; such an action would be without any design to it: it would be working without doing anything, *agendo nihil agere*."[13]

In "Real Quantities and Their Sensible Measures," Lawrence Sklar sets that problem to one side. He goes back to the Scholium and puzzles over Newton's attempts to provide a unified treatment of the various "real quantities"—absolute space, time, and motion. He points out that Newton advances very different kinds of arguments for absolute time than for absolute space and motion. Absolute time, says Newton, is presupposed in our practice of correcting our "sensible measures" of time; absolute space and motion, by contrast, are argued for from dynamical considerations—namely, the dynamical effects of absolute rotations.[14] Sklar notes that interpreters of Newton have sometimes emphasized one set of considerations to the neglect of the other, and suggests that this difference in emphasis may lie behind some of the contemporary debate between realists and representationalists on the nature of space-time. According to Phillip Bricker, however, the distinctions Sklar draws between the problems of absolute time, space, and motion do not go to the heart of the debate between realists and representationalists. In "Absolute Time versus Absolute Motion" he sets out to clarify the ways in which these problems differ from one another, especially with regard to the question: How do absolute time, space, and motion differ with respect to attributions of causation?

v

As the passage I quoted earlier shows, Leibniz is vigorously opposed to the idea that there is an absolute space. His own account of space and time is relational.[15] In his correspondence with Clarke he writes: "I hold space to be something merely relative, as time is; . . . I hold it to be an order of coexistence, as time is an order of successions."[16] In saying this he is rejecting the substantial view of space he finds implicit in Newton's account; that is, he rejects the view that space is some kind of *thing*, "a real absolute being" that has attributes like eternity and immensity.

From the Newtonian side, Clarke responds that it is no part of Newton's position that space is a substance: "Space is not a being, an eternal and infinite being, but a property, or a consequence of the existence of a being infinite and eternal."[17]

For all parties, the nature of space and the nature of the Deity are closely related issues. J. E. McGuire and John Carriero, writing in this volume, differ over what Newton's views are on the relation in question. At issue between them is whether Newton regards infinite space and infinite time as attributes of God himself, or as causally dependent on him. Very roughly, McGuire embraces the first alternative, Carriero the second. In "Predicates of Pure Existence," McGuire argues that, for Newton, infinite extension and infinite duration are not essential attributes of God, but attach to his actuality as general conditions of his existence; to the extent that Newton thinks of space and time as causally dependent on God, the causality involved is Henry More's "emanative causality,"[18] and a causal dependency of that kind is barely distinct from an ontic dependency. In response Carriero contends that McGuire shortchanges Newton's insistence that space and time are "something." According to Carriero, Newton takes God to be the efficient cause of space and time, thus presupposing that they are beings distinct from God, and making otiose an unwelcome distinction between his essence and his existence.

vi

With hindsight we can see that a science of Galilean primary qualities, all expressible in terms of space and time, will give us a kinematics but not a dynamics; Newtonian dynamics requires the additional concepts of *mass* and *force*.[19] More generally, while remaining within a mathematical representation of the physical world, seventeenth-century philosophy needed an explanation of change. From these explanations Aristotelian formal and final causes were to be rigorously excluded.[20]

According to the Galilean principle of inertia, taken over by Descartes, the changes to be explained were not changes in place but changes in natural motion.[21] Descartes made the further claim that any change in the motion of a body was the result of a collision with other bodies.[22] It followed that only contact forces could be appealed to in scientific explanations, and in Part IV of the *Principles* we see Descartes duly postulating, or "hypothesizing," interactions of this kind between unobservably small particles to account for such phenomena as the fragility of glass and the action of a magnet.[23] His view became the orthodox view within mechanical philosophy, endorsed by Boyle, by Huygens (hence his interest in the rules of collision), and by Leibniz, by atomists and by nonatomists alike.

By appearing to postulate action at a distance, Newton's theory of gravity ran directly counter to orthodoxy. It was for that reason violently attacked, witness Leibniz's dismissive comment: "Tis also a supernatural thing, that bodies should attract one another at a distance, without any intermediate means . . . For these effects cannot be explained by the nature of things."[24] Newton was accused of reintroducing into philosophy "the occult qualities of the schools"— mysterious powers or virtues which simple substances had within themselves.[25]

His response to this was to challenge the Cartesian program, specifically Descartes' "method of hypothesis."[26] His challenge had two components. In the first place he rejected the requirement that all scientific explanation be in terms of contact forces; the formulation of mathematical laws governing a set of phenomena is itself, he says, a valuable contribution to science. This allowed him to remain officially neutral on the question of action at a distance. His words are well known: "To derive two or three general principles of motion from phaenomena, and afterwards to tell us how the properties and actions of all corporeal things follow from these manifest principles, would be a very great step in philosophy, though the causes of those principles were not yet discover'd."[27]

Second, he attacked the "method of hypothesis" itself, on the grounds that any number of incompatible mechanical explanations could be given for a phenomenon (all, since they involve unobservable particles, equally problematic). Thus: "If anyone offers conjectures about the nature of things from the mere possibility of hypotheses, I do not see how anything certain can be determined in any science; for it is always possible to contrive hypotheses, one after another, which are found rich in new tribulations."[28]

But now consider Query 29 of the *Opticks* (p. 370), which begins, "Are not the rays of light very small bodies emitted from shining substances?" This is surely a conjecture about the truth of things which goes beyond any inductively grounded generalization. In "Newton's Corpuscular Query and Experimental Philosophy" Peter

Achinstein looks at this Query to see whether Newton does indeed violate his own methodological prescriptions. He provides a detailed analysis of Newton's use of key terms such as *phenomenon*, the *deduction* of propositions from phenomena, and *hypothesis*, and examines Newton's presentation of the corpuscular theory of light. He then suggests that we may plausibly think of the "corpuscular hypothesis" as a proposition to which Newton is making a "weak" rather than a "strong" inference. That is to say, the hypothesis is not, for Newton, a proposition deduced from phenomena in accordance with his own Rules of Reasoning in Philosophy; its explanatory success, however, together with the inadequacies of its only competitor, the wave theory, do provide some evidence for its truth. As I say in my comments ("Reason and Experiment in Newton's *Opticks*"), I think that Achinstein's "weak inference" involves an induction that Newton would be loath to make, and, more generally, that he neglects the importance Newton attaches to organizing the propositions of science within a deductive framework; to use Newton's words, that he emphasizes *analysis* at the expense of *synthesis*.[29]

One way to reconcile some, at least, of Newton's statements about hypotheses with the role they played in his thought is to say that he regarded them as suggestive guides to research, but as having no place in a formally presented science. His treatment of the vexed problem of gravity seems to bear this out. An early speculation was that gravity might be caused by "an aetherial spirit," and a possible mode of action of this spirit is described in a letter to Boyle of 1678.[30] In the *Principia*, however, it is not until the last paragraph of the final Scholium that this speculation is aired, and there, after a very brief discussion, Newton closes by saying: "These are things that cannot be explained in a few words, nor are we so furnished with that sufficiency of experiments which is required to an accurate determination and demonstration of the laws by which this electric and elastic spirit operates" (p. 547).

vii

However slender the possibility of explaining gravitational attraction, on one issue Newton is quite definite: he does not consider gravity to be "essential and inherent in matter."[31] In the *Metaphysical Foundations of Natural Science*, written a century after the *Principia*'s publication, Immanuel Kant claimed that here Newton is "at variance with himself." Kant's reasons for saying so are the topic of Michael Friedman's paper, "Kant and Newton: Why Gravity Is Essential to Matter." Friedman suggests that Kant's argument is of a kind familiar to readers of the *Critique of Pure Reason*; it is an account of how matter is possible as an object of knowledge and experience. As Friedman reconstructs it, this involves (a) showing how a *measure* of mass becomes available to us, and (b) giving objective content to the idea of matter as the movable in space. To show how these two projects are linked, Kant—or Friedman acting on Kant's behalf—turns to Newton. From the *Principia* we see how, by assuming the law of gravitation, we can compute the relative masses of the sun and planets, and hence find the position of the center of mass of the solar system. This will not itself give us an absolute space against which the absolute motion of matter can be defined, since the solar system is itself in orbit with respect to the Milky Way, which in turn moves with respect to other galaxies. It does, however, offer a frame of reference against which "real" motion can be approximately determined; real motion can then be seen as a regulative idea to which successive approximations guide us. Thus, on the assumption of the universality of the gravitational law, requirements (a) and (b) are both satisfied. Gravity is seen as essential to matter, in the sense of being required for our cognizing of it.

In response Robert DiSalle, in "The 'Essential Properties' of Matter, Space, and Time," raises three problems. First, he asks whether Kant displays as deep an understanding of the role of absolute space in Newtonian physics as does Newton. Second, he points out that, although the only feasible way to compare planetary masses may be through their gravitational interactions, this need not *in principle* be the only way to do so, given Newton's three laws of motion. Third, he claims that some of the structure of Newtonian space-time is presupposed in the "passive principles" of Newton's laws of motion, and that these in turn are presupposed in the "active principle" of gravitation, a chain of conceptual dependence which is reflected in the ordering of the *Principia*. On Friedman's reading of Kant, this chain would become a circle.

I do not propose to adjudicate this debate. Friedman's paper and DiSalle's responses show us, however, that Kant's *Metaphysical Foundations of Natural Science* deserves more attention, from students of Kant and of Newton alike, than it has yet received. Within it we can see Newton's thought refracted, as it were, through the prism of Kant's critical philosophy, and thereby learn more about both.

viii

The final pair of papers—"Ethical Implications of Newtonian Science" by Errol Harris and "Modern Ethical Theory and Newtonian

Science" by Philip Grier—extend the discussion to the present day and look at the ethical implications of Newton's work. Harris presents a historical analysis to show that Newtonian science has given rise to patterns of ethical theory that (a) are entirely concerned with human agents and their desires, (b) take an instrumental attitude to the natural world, and (c) inevitably draw us to emotivism and relativism. In his closing remarks Harris also outlines the possibility of a kind of ethical theory that, by responding to new holistic patterns of thought within contemporary science, could be properly sensitive to the relation between human beings and their environment.

It is worth noting that, Newton's talk of "solid, massy, hard, impenetrable and moveable particles" notwithstanding, holistic considerations were not foreign to his thought. This should be scarcely surprising; after all, he originated the revolutionary proposal that every part of the material world is in continuous dynamical interaction with every other part. Here is an example from his first letter to Bentley:

> The planets of Jupiter and Saturn . . . are vastly greater and contain a far greater quantity of matter [than the rest]; which qualifications surely arose, not from their being placed at so great a distance from the sun, but were rather the cause why the Creator placed them at great distance. For, by their gravitating powers, they disturb one another's motions very sensibly, . . . and had they been placed much nearer to the sun and to one another, they would, by the same powers, have caused a considerable disturbance in the whole system.[32]

Newton here pictures the solar system as a dynamic system in a delicately balanced state of internal equilibrium. No doubt the theological cast of such arguments has made them unappealing, and that is one reason why we have inherited less from Newton than he bequeathed to us. As we strive to repossess it, Harris' constructive project merits our serious attention. Grier perceptively describes it as a search for a new ethics of "natural law"; in his essay he supplements the historical account given by Harris, drawing particular attention to the way ethical theory has been shaped not only by Newtonian science but also by Romanticism and by the philosophical anthropology associated with it.

Harris' paper prompts the general question: To what extent can we identify what is often described as the "Newtonian world-view" with Newton's own ideas? Since the anniversary that this volume of papers celebrates seems an apt occasion to discuss the relation between the two, I will spend the remaining pages of this introduction doing so. The phrase "the Newtonian world-view" ("Newtonianism" for short) is rather vague; it is a title bestowed on a cluster of attitudes thought to be entailed by, or at least historically associated with, Newton's science. Fritjof Capra, for example, assumes in his book *The Turning Point* that there is an identifiable and historically important set of attitudes of this kind, and his characterization of them would be widely accepted;[33] for specificity, this is the characterization I shall use. Suggestively enough, Chapter 2 of his book is entitled "The Newtonian World-Machine."

Newtonianism contains what may be called the pure science set out in the *Principia* and the *Opticks*. To be sure, in the two centuries after their publication aspects of this science were amended; but, to answer a question I raised earlier in this essay, the concepts of "Newtonian mechanics" are still recognizably Newton's. Newtonianism also takes over Newton's epistemological assumption that science can be functionally autonomous; it is not beholden, as it were, to other branches of philosopohy or theology.[34]

But acceptance of Newton's science and of its functional autonomy is consistent with many different kinds of attitudes concerning ethics, theology, and the natural world. It is consistent, for example, with the view that the material world is a world from which God has been banished, and within which judgments of value are irredeemably subjective, and views of this kind are typically thought of as "Newtonian." But it does not entail those views. Nor, interestingly, did it prompt their historical emergence in any immediate way.

With regard to ethics, for example, although there were eighteenth-century relativists (Montesquieu comes to mind), they were not Newtonians. Indeed, it was precisely those eighteenth-century philosophers who paid most attention to Newton who produced moral theories, albeit of very different kinds, that were both objectivist and universalist.[35] Thus Thomas Reid thought that moral intuition enabled us to discern objective moral principles;[36] Kant's moral imperatives held for all rational agents; and, according to Hume, ethical judgments acquired their objective character by being grounded in a universal human response. His Second Enquiry, surely the most self-consciously Newtonian book on ethics ever written, argues that the *principle of humanity* is as much an objective and experimentally confirmed fact as Newton's law of gravitation, to which Hume frequently compared it.[37]

Again, from a twentieth-century perspective, the Newtonian world may well seem to be one from which God is absent; but for Newton and many of his contemporaries, the order and harmony of the universe revealed by Newton's science bore continual testimony to the power and presence of God. In the opening sentence of his first letter to Bentley, Newton wrote: "When I wrote my treatise about our system, I had an eye upon such principles as might work with considering men for the belief in a Deity; and nothing can rejoice me more than to find it useful for that purpose."[38] Its usefulness is attested by the way in which the Deist movement of the next century greeted both his science and the argument from design that he drew from it. Addison's splendid hymn, for example, wholly inspired by the *Principia*, sings of the sun and moon, the planets, and the stars, and tells us that

> In Reason's ear they all rejoice,
> And utter forth a glorious voice,
> Forever singing as they shine,
> "The Hand that made us is divine."

On the question of God's immanence in the world, Newton may have said—in fact he did say—that he "governs all things, not as the soul of the world, but as Lord over all" (p. 544). But before this leads us to agree with Capra that, in Newton's world, "the physical phenomena themselves were not considered to be divine in any sense,"[39] we need to reconcile it with what he says a page later: "God is the same God, always and everywhere, He is omnipresent not *virtually* only, but also *substantially*, for virtue cannot exist without substance" (p. 545). The precise sense in which Newton thought of God as *omnipresent* is, as I have already mentioned, the subject of J. E. McGuire's paper in this volume.

Newton's science is often associated, via Newtonianism, with the advent of technology and with the prevalence of instrumental attitudes toward the natural world. It is true that, a century after Newton's time, there occurred the Industrial Revolution, and that, some time later, the technological possibilities of science were realized. But this does not imply that "with the rise of Newtonian science nature became a mechanical system that could be manipulated and exploited."[40] Here Newton's name is being attached to a specifically Baconian project. His own attitudes toward the material world are eloquently expressed in this passage from the *Opticks*.

> The main business of natural philosophy is to argue from phenomena without feigning hypotheses, and to deduce causes from effects, till we come to the very first cause, which certainly is not mechanical; and not only to unfold the mechanism of the world, but chiefly to resolve these and such like questions . . . Whence is it that Nature does nothing in vain; and whence arises

all that order and beauty which we see in the world? To what end are comets, and whence is it that planets move all one and the same way in orbs concentrick, while comets move all manner of ways in orbs very excentrick; and what hinders the fixed stars from falling upon one another? How came the bodies of animals to be contrived with so much art, and for what ends were their several parts? Was the eye contrived without skill in opticks, and the ear without knowledge of sounds?[41]

Much hard deconstructive work would have to be done to show that these words express an instrumental or an exploitative attitude to the natural order. That in turn suggests that there is no necessary connection, logical or psychological, between such attitudes and Newton's science, and hence that, if this science has been appropriated by an ideology that seeks to exploit the natural world, then the explanation for this must lie outside the internal dynamics of intellectual history.

In sum, granted that there is a general world view of the kind that Capra describes, by calling it Newtonian we risk mistaking both its dominant characteristics and its historical provenance. (Plausibly the place to look for these might be within prevalent modes of social and economic organization.) As for Newton, he declines to be classified as a "Newtonian," largely because he resists classification. As the papers in this volume show, although we may find the antecedents, or trace the influence, of various elements of his thought, we cannot squeeze him without remainder into any comfortable philosophical category. After three hundred years he remains a man of singular genius.

Notes

1. Sir Isaac Newton, *Philosophiae Naturalis Principia Mathematica*, trans. A. Motte, rev. F. Cajori (Berkeley: University of California Press, 1934), II, 507–522. Subsequent page references to the *Principia* will be given in the text.
2. E. N. da C. Andrade, "Newton and the Science of his Age," in *Sir Isaac Newton, President of the Royal Society, 1703–1727: Lectures Delivered on the Occasion of the Anniversary Meeting of the Royal Society to Commemorate the Tercentenary of His Birth, 1942* (London: The Royal Society, 1942), p. 15.
3. E. A. Burtt, *The Metaphysical Foundations of Modern Science*, 2d ed. (New York: Doubleday Anchor, 1954), preface.
4. Burtt, p. 309.
5. See "Newton and Descartes," in A. Koyré, *Newtonian Studies* (Chicago: University of Chicago Press, 1965).
6. René Descartes, *Principles of Philosophy*, trans. V. R. and R. P. Miller (Dordrecht: Reidel, 1983); it was originally published in 1644. To avoid confusion, from now on I will use its English title or just the abbreviation the *Principles*. Page references will be given in the text.

7. Sir Isaac Newton, *Opticks*, 4th ed., ed. I. B. Cohen (New York: Dover Books, 1952), p. 1.

8. In a draft preface intended for the *Opticks* but not used, God is among the "principles of philosophy" that Newton discusses; much of what he says here is later incorporated into Query 28 of the second edition of the *Opticks*. The text of the draft preface is given in J. E. McGuire, "Newton's Principles of Philosophy: An Intended Preface for the *Opticks* and a Related Draft Fragment," *British Journal for the History of Science* 5 (1970), 178–182.

9. *Opticks*, Query 28, p. 370.

10. Galileo Galilei, "The Assayer," in *Discoveries and Opinions of Galileo*, trans. and ed. Stillman Drake (New York: Doubleday, 1957), p. 238.

11. Ibid., p. 274.

12. For Newton's views on the question of primary qualities, see Howard Stein's paper in this volume.

13. Leibniz's Fifth Paper, in *The Leibniz-Clarke Correspondence*, ed. H. G. Alexander (Manchester: Manchester University Press, 1956), p. 63. Against Leibniz, Julian Barbour points out the role that absolute space plays—and was known by Newton to play—in the very definition of uniform linear motion; see Julian Barbour, *The Discovery of Dynamics* (Cambridge: Cambridge University Press, in press).

14. But see Ronald Laymon, "Newton's Bucket Experiment," *Journal of the History of Philosophy* 16 (1978), 399–413.

15. For the various senses in which space and time can be "absolute," see Michael Friedman, *Foundations of Space-Time Theories* (Princeton: Princeton University Press, 1983), pp. 62–70.

16. Leibniz's Third Paper, *Leibniz-Clarke Correspondence*, pp. 25–26.

17. Clarke's Third Reply, ibid., p. 31.

18. Newton's views on space and time were deeply influenced by the Cambridge Platonist Henry More (1611–1687). See Alexander Koyré, *From the Closed World to the Infinite Universe* (Baltimore: Johns Hopkins University Press, 1957), esp. chaps. 6 and 7.

19. Thus, when Newton speculates (*Opticks*, p. 400) that "God in the beginning form'd matter in solid, massy, hard, impenetrable, moveable particles, of such sizes and figures, and with such other properties and in such proportion to space, as most conduc'd to the end for which He form'd them," the inclusion of the attributes *massy* and *hard* is significant.

20. At a certain level, however, Newton reinstates final causes, as in the last clause of the quotation in n. 19. See also the first letter to Bentley, quoted in part in the last section of this essay and available in H. S. Thayer, ed., *Newton's Philosophy of Nature: Selections from His Writings* (New York, Hafner, 1953), pp. 46–50.

21. *Principles* II, §37, p. 59. Galileo, however, identifies natural motion with uniform circular motion, whereas Descartes identifies it with uniform rectilinear motion (*Principles* II, §39, p. 60).

22. Ibid., §40, p. 62.

23. Indeed, Descartes tells us that "no phenomena of nature have been omitted in this treatise" (*Principles* IV, §99, p. 282).

24. Leibniz's Fourth Paper, *Leibniz-Clarke Correspondence*, p. 43.

25. Leibniz's Fifth Paper, ibid., p. 92.

26. For an account of this debate and its continuation through the eighteenth and early nineteenth centuries, see Larry Laudan, *Science and Hypothesis* (Dordrecht: Reidel, 1981).

27. *Opticks*, pp. 401–402.

28. Letter to Oldenburg, July 6, 1672, quoted by Cajori in the appendix to the *Principia*, p. 673. Note that this objection actually precedes the dispute over gravity.

29. See *Opticks*, pp. 404–405.

30. Letter to Boyle, February 28, 1678, in Thayer, pp. 112–116; Newton returns to those topics in Query 21 of the *Opticks*.

31. Second lettter to Bentley, January 17, 1693, in Thayer, p. 53; see also the advertisement to the second edition of the *Opticks* (p. cxxiii).

32. First letter to Bentley, December 10, 1693, in Thayer, p. 49.

33. F. Capra, *The Turning Point* (New York: Bantam Books, 1983).

34. This attitude first finds expression (I think) in Galileo's "Letter to the Grand Duchess Christina"; see Drake, esp. p. 183.

35. I use Grier's terminology; see his footnotes 7, 8, and 9.

36. For Reid's Newtonianism, see Laudan, chap. 7.

37. See R. I. G. Hughes, "Hume's Second Enquiry: Ethics as Natural Science," *History of Philosophy Quarterly* 2 (1985), 291–307.

38. First letter to Bentley, December 10, 1692, in Thayer, p. 46.

39. Capra, p. 66.

40. Ibid., p. 40.

41. *Opticks*, pp. 369–370.

Chapter 2

On Locke, "the Great Huygenius, and the incomparable Mr. Newton"

Howard Stein

The quoted phrase occurs in the Epistle to the Reader, prefatory to Locke's great *Essay*: imagining himself "censured" for "pretend[ing] to instruct this our knowing age," Locke acknowledges that such a purpose is indeed the only rational justification for publishing such a book, and proceeds to indicate the place he claims to fill in "the Commonwealth of Learning": "Everyone," he says, in a nicely gauged crescendo of admiration, "must not hope to be a Boyle, or a Sydenham; and in an Age that produces such Masters, as the Great Huygenius, and the incomparable Mr. Newton, . . . 'tis Ambition enough to be employed as an Under-labourer in clearing Ground a little, and removing some of the Rubbish, that lies in the way to Knowledge."

In recent years Locke's interest (and involvement) in the science of his time, and the influence of that science on his philosophy, have been the subject of much discussion. This seems to me a salutary historiographic development, and I should be pleased to have some of what I shall say here count as a contribution to that discussion; but my main concern is a bit different. I believe (and this, at least in our century, has not been a common opinion) that in the work of clearing the underbrush and removing obstacles to the advancement of knowledge—just as in the positive work of the advancement itself—Newton was nonpareil; and I wish to illustrate this claim, in part, by a comparison with Locke. But to prevent misunderstanding, let me emphasize at the outset that, notwithstanding Locke's characterization of this task of an "underlaborer," what we have to do with is nothing less than epistemology—or methodology—and metaphysics; my contention is that Newton attained, and deployed, in most intimate connection with his scientific work, conceptions both of method and metaphysics of a subtlety that has not been generally appreciated—conceptions from which there is still something to learn.

Besides the three mid-to-late seventeenth-century figures I have mentioned, two from the early part of the century are of great relevance to my subject: Descartes and Galileo. Huygens is sometimes described as a Cartesian; and, to be sure, he was—in his student years. In 1650, on hearing the news of Descartes's death in Sweden, the twenty-one-year-old Huygens wrote a verse elegy, whose concluding stanza is as follows:

> Nature, put on mourning—come and be first mourner
> For the Great Descartes, and show your despair;
> When he went into darkness, you lost your light—
> It is only by this torch that we have been able to see you.[1]

But his riper opinion, written some forty-odd years later, is very different, and deserves to be considered. In 1693, while making notes on Baillet's life of Descartes (chiefly to correct errors of fact concerning the Huygens family), Huygens comes to a passage referring to Descartes's poor opinion of the explanation offered by Gassendi of the phenomenon of "parhelia"; and is led to a disquisition several pages long on the development of modern natural philosophy. I quote from it:

> He rightly disdains the explanation of parhelia by M. Gassendi, which is mistaken; but the one he himself gives in his Meteors is ridiculous and very easy to refute . . .
> M. des Cartes had found the way to make his conjectures and fictions pass for truths. And those who read his Principles of Philosophy experienced somewhat the same as do the readers of Romances, which give pleasure and make the same impression as veritable histories . . . It seemed to me when I first read this book of the Principles that everything in it was as good as could be [que tout alloit le mieux du monde], and I believed, when I found some difficulty there, that it was my fault for not correctly understanding his thought. I was only 15 or 16 years old. But having since then discovered there from time to time things visibly false, and others highly improbable, I have retreated far from the predilection I then had, and at the present hour I find almost nothing that I can approve as true in all his physics, or metaphysics, or meteors.[2]

Descartes's physics is indeed bizarre. It is hard to understand how anyone who has actually read Descartes's *Principia* could believe (what is often asserted) that Descartes's program involved "the reduction of physics to geometry," or alternatively could claim, as a historian who really should know better has done, that "[the] form [of Descartes's physics] is precisely that of Newton's."[3] To be sure,

Descartes does say that he "neither accepts nor desires any other principles in Physics, than in Geometry or in abstract Mathematics";[4] and of course he maintains that the one "principal attribute" constituting the "nature of corporeal substance" is "extension in length, breadth, and depth";[5] but one will search long and hard to find anything like mathematical reasoning in the book.

What comes closest, there, to a principled foundation for mathematical arguments concerning the operations of nature is the set of rules Descartes gives "to determine by how much any body's motion is changed by coming into contact with other bodies."[6] In view of his insistence that contact with other bodies is the *only* circumstance occasioning changes of motion, such rules ought to form the basis for the theory of motion; and in view of his insistence that all the "diversity of forms" in matter is based upon motion,[7] these rules should in fact constitute the foundations of physics.

Now, the two Cartesian principles I have just mentioned (that all "diversity of forms" in matter fundamentally consists in arrangements of motions, and that the laws of motion are the laws of *change* of motion *through contact*) are the essential principles of what became known as the *mechanical*, or—under the designation introduced by Boyle—the *corpuscularian* philosophy. When this general view takes the more particular (and anti-Cartesian) form in which it is supposed that all matter is constituted out of ultimate, discrete, *indivisible* corpuscles—"atoms"—it is clear that there is entailed a definite program for fundamental explanation in physics: one must know the characteristics of the atoms; one must know their arrangement and motions in the particular systems of interest; and one must know the general laws of the communication of motion through contact—that is, the laws of impact.

But Descartes's seven rules of impact are thoroughly absurd;[8] and Descartes is driven to defend himself against the objection that these rules are in blatant conflict with experience, by maintaining that the rules hold only for two bodies colliding *in vacuo*, whereas all observed impacts occur within an ambient medium.[9] This is a remarkable defense: according to Descartes, the notion of two separated bodies *in vacuo* is a formal contradiction; and it is a rather odd thing to see particular predictions made of the various results to be expected under various self-contradictory conditions. In any case, Descartes himself not only offers an excuse for his rules—although at the same time maintaining their absolute certainty[10]—but also suggests that the reader may safely ignore them, since they are not needed to understand the rest of the work.[11] (Imagine telling a reader of Newton's *Principia* that the laws of motion are unnecessary for an understanding of the rest of the book!—so much for the "identity of form" of the two.)

It was clearly this farrago of incoherencies that first led Huygens to moderate his admiration of Descartes; indeed, within a year or two of his composition of the elegy from which I have quoted, Huygens was in full possession of the correct laws of the elastic central impact of nonrotating spheres. In the course of his critical remarks of 1693, he refers to Descartes's "laws of motion in colliding bodies, which"— Huygens says—"he thought to make pass for true by giving it out that all of his physics would be false if these laws were false." (It is well known that Descartes did make such a statement about his theorem that light is propagated instantaneously;[12] I know of no authority other than that of Huygens for a similar assertion about the rules of impact.) Huygens' comment on this mode of rhetoric is, "This is almost as if he wished to prove them by taking an oath."

I want now to describe briefly Huygens' two fundamental investigations bearing on the general laws of motion: that of impact (extending from 1652 to 1667), and that of centrifugal force (dating from 1659).

In the critical notes on Descartes already twice cited, Huygens gives a brief appreciation of anti-Aristotelian natural philosophy among both the ancients (Democritus, Epicurus, "and many others")—who, however, he says, "did not explain a single phenomenon in a satisfactory way"—and the moderns. Among the latter "Telesius, Campanella, Gilbert retained . . . many occult qualities, and had not sufficient inventiveness nor sufficient mathematics . . . ; no more did Gassendi . . . Verulamius [that is, Bacon] . . . taught very good methods for building a better [philosophy] by making experiments and putting them to good use . . . But . . . he did not understand Mathematics and lacked penetration for the things of physics." Only one figure comes in for unqualified praise: "Galileo had, in point of genius and of knowledge of Mathematics, all that is needed to make progress in Physics, and . . . he was the first to make beautiful discoveries concerning the nature of motion, although he left a great deal to be done. He did not have so much boldness or presumption as to pretend to explain all natural causes, nor the vanity to wish to be the head of a sect [as Descartes did]. He was modest and loved truth too much; moreover, he believed he had acquired sufficient reputation— and one that would endure forever—by his novel discoveries." Huygens, in effect, regards himself as the corrector of Descartes, and the continuator of Galileo. The results of Galileo are fundamental to both the investigations of Huygens that I have named.

Huygens based his general theory of impact on three principles: that of inertia,[13] that of Galilean relativity,[14] and the principle that "by a motion of bodies that results from their gravity, their common center of gravity cannot be raised."[15] That this last follows from the impossibility of "perpetual motion" Huygens well knew.[16] As to the principle of relativity, he had of course encountered it, in rough but pregnant form, in Galileo; and for Huygens himself it was associated with deep philosophical convictions about the nature of motion (the fact that Descartes's rules of impact violate this principle—although Descartes professes a quasi-relativistic theory of the nature of motion—was surely one of the defects Huygens early discovered in those rules).

In addition to these three general principles, Huygens relies crucially on Galileo's laws of falling bodies: (1) that the speed acquired by a body, in frictionless fall *in vacuo* through a height *h*—whether vertically, or along an incline, or under any other frictionless constraint— is the same for all bodies, and depends only on *h*: is in fact proportional to the square root of *h*; and (2) that the speed resulting from a fall through the height *h* is able—along any frictionless upward path *in vacuo*—to lift the body to the same height *h*.

With the help of these assumptions, Huygens succeeded— apparently in 1667[17]—in proving a very general result about the "encounter" of arbitrarily many bodies, of any nature whatever—elastic, inelastic, or partially elastic (Huygens' terms are "hard," "soft," and "semiresilient")—provided, at least, that no rotations are involved. The gist of his beautiful argument, in the case of two bodies and direct central impact, is the following: Let the interacting bodies be called A and B, and let A and B likewise denote their "sizes." Let their velocities before the interaction—reckoned as positive toward one side, negative toward the other—be u_A and u_B; those after the interaction, u'_A and u'_B. Then if g is the acceleration in free fall, the initial velocities can have been acquired in falling through heights $h_A = (u_A)^2/2g$, $h_B = (u_B)^2/2g$ respectively; and in the initial resting configuration of the bodies, the height of their center of gravity (above the level on which they come to interact) will have been $(Ah_A + Bh_B)/(A + B)$. Again, if the resulting velocities are used to lift the bodies, the heights they can attain will be $h'_A = (u'_A)^2/2g$ and $h'_B = (u'_B)^2/2g$, and the height of the center of gravity $(Ah'_A + Bh'_B)/(A + B)$. By Huygens' assumption, therefore, we must have: $Ah'_A + Bh'_B \le Ah_A + Bh_B$; in other words, $A(u'_A)^2 + B(u'_B)^2 \le A(u_A)^2 + B(u_B)^2$.

Note that we have, so far, used only Galileo's laws and the impossibility of the center of gravity rising above its initial (resting) value. Note too that we have made no special reference to *impact* (except for the presumption that all "fundamental" interaction *is* impact): the argument so far applies to *any* "irrotational" encounter. On the other hand, the result, having the form of an inequality, is rather weak. Huygens' really beautiful discovery was that this inequality can be made to yield a fundamentally important equation—"a wonderful law of Nature," he calls it ("une loy admirable de la Nature")[18]—by applying to it the principle of relativity. For let the same interaction be viewed by an observer moving with velocity v; to this observer the initial and final velocities of the bodies will appear to be $u_A - v$, $u_B - v$, $u'_A - v$, $u'_B - v$, respectively; and since the same laws must hold for the moving observer[19] as for one at rest, we must have:

$$A(u'_A - v)^2 + B(u'_B - v)^2 \leq A(u_A - v)^2 + B(u_B - v)^2;$$

i.e., after expanding the squares, canceling the terms $Av^2 + Bv^2$, which occur on both sides, and moving all terms involving v to one side, all others to the other:

$$2v(Au_A + Bu_B - Au'_A - Bu'_B) \leq A(u_A)^2 + B(u_B)^2 - A(u'_A)^2 - B(u'_B)^2$$

Now, the term on the right does not depend on v; and v can be taken to be positive or negative, and as large as we wish. It follows that this inequality will surely be violated for some choice of v, unless the expression in parentheses on the left is equal to zero.[20] So we have established that, under the very general conditions specified, $Au'_A + Bu'_B = Au_A + Bu_B$. Moreover, there is no difficulty at all in extending the argument to the case of n bodies, $A_1, \ldots, A_n$, with initial (vectorial) velocities $\bar{u}_1, \ldots, \bar{u}_n$, and final velocities $\bar{u}'_1, \ldots, \bar{u}'_n$, obtaining the result:

$$\Sigma A_i \bar{u}'_i = \Sigma A_i \bar{u}_i$$

—that is, the general principle of the conservation of momentum. (To complete the theory of the direct impact of two bodies, Huygens gives a supplementary relation which, with the former, suffices to determine u'_A and u'_B, the other quantities being given. For "hard" bodies, this relation is: $u'_A - u'_B = u_B - u_A$—the bodies separate with the same relative speed with which they approached;[21] for "soft" bodies: $u'_A = u'_B$—the bodies remain in contact. The corresponding supplementary relation for partially elastic bodies was discovered experimentally by Newton, and first announced in the *Principia*.[22])

Huygens' investigation of centrifugal force has a rather different character, since here the essential difficulty lay in finding the "appropriate" *definition*—that is, *concept*—of the "quantity" of centrifugal force. As in the case of impact, he finds a strategy that leads with utmost clarity and simplicity to the desired end—particularly remarkable here, since the notion of an "appropriate" definition is not itself a well-defined one.

The idea of Huygens' argument is this:[23] First, we may take as the paradigm phenomenon the tug exerted upon a string by a body held by that string on a uniformly rotating disk; we want to know "how strong a tug." Next, we reflect that a weight, hung vertically by a string, also exerts a tug, whose "strength" or "intensity" we may reasonably estimate by the magnitude of that weight; so we are led to ask whether some kind of equivalence can be established between these two sorts of phenomena. Now, Huygens remarks, weight is a "tendency" to *fall*; and, as Galileo has shown, and Riccioli and Huygens himself have confirmed by careful experiments, freely falling bodies—and also bodies that descend along inclined planes—move with a definite and uniform acceleration, the same for all bodies (in the vertical direction, or along a given incline; Huygens expresses this result not in terms of acceleration, but by the equivalent proposition that the distances covered in successive equal time intervals starting from rest are in the proportion of the successive odd numbers starting from unity). To be sure, this statement requires some amendment for bodies falling in air, as Huygens notes at once; but he adds that he needs, for his argument, only the fact that the proposition holds *in the limit of "arbitrarily small spaces"*—which in effect means that the acceleration at the instant of release from rest is nonzero, and the same for all bodies.[24]

The difference of acceleration in vertical fall and along an incline plays a critical role in the argument; more especially, the fact that the acceleration when the body is released, and the tug it exerts on the string when it is restrained, are both diminished—and, Huygens says, in the same proportion—when the inclination of the path is reduced.[25] Finally, to show that only the *initial* acceleration matters, Huygens asks us to consider a weight suspended in contact with a *curved*, rather than a plane, guiding surface; and in particular, one whose direction at the point of contact is vertical. On the one hand, no diminution of "apparent weight"—of "tug"—is experienced in this case. On the other hand, when the body is released, both its direction of motion and the magnitude of its acceleration vary continuously; but its initial acceleration is vertically downwards, and equal to that in free fall.

We now turn to the body held on the uniformly rotating disk. If released, it will (we here ignore gravity as well as air resistance) move along the line tangent, at the point of release, to the circle in which it had been going, at a constant speed, equal to the circumferential velocity of the rotation. *How will this look to an observer on the disk?* The answer is simple and elegant: the body will move, relative to the disk, along the curve that would be described by the endpoint of a thread, held taut, being unrolled from the disk as from a spool—and will move just as if the thread were being unwound at a uniform rate.[26] This curve—the *involute* of the circle—is easily seen to be perpendicular to the circle at the initial point of unwinding; and the body's motion along that curve is, at the initial instant, an accelerated motion starting from rest. Huygens proceeds to evaluate this initial acceleration; without expressing the result explicitly in this form, he shows in effect that if the radius is r and the circumferential velocity v, the initial acceleration relative to the observer on the disk will be v^2/r. So for the rotating observer—or for the string—the *restrained* body is restrained from yielding to its tendency to accelerate radially outward to that degree. "But," he says (implicitly invoking what Clifford Truesdell and his associates have called "the principle of material frame-independence"), "this tendency is entirely similar to that by which heavy bodies suspended from a thread strain to go downwards";[27] and he concludes that the degree of this "straining"—that is, the centrifugal force—is, like the force of the weight, proportional both to the acceleration striven for and to the "size" of the body.

Thus Huygens' argument leads to a result that can be immediately generalized as follows: If, from the point of view of an observer (in no matter what state of motion), a body of "size" or "mass" m is held stationary by a string; and if, in the same conditions, but released from the string, the body would move (relative to that observer) with initial acceleration a in a given direction, then the tug on the string holding the body will be in that direction, and of a magnitude proportional to ma.[28]

I have said earlier that a consequence of Huygens' discussion of *interactions* of bodies is the general principle of the conservation of momentum. The result just formulated looks temptingly like Newton's second law of motion. Can we say that the essential contents of the introductory material of Newton's *Principia*—the "Definitions" and the "Laws of Motion"—were in Huygens' possession many years before Newton published them?

The answer, I think, is not entirely simple—not an unqualified yes or no. Newton himself, it should be emphasized, did *not* claim originality for the definitions and laws: he says, "Hitherto I have laid down such principles as have been receiv'd by Mathematicians, and are confirm'd by abundance of experiments"[29] (and refers in particular to Galileo, Sir Christopher Wren, John Wallis, and Huygens). But, as I shall explain presently, I believe that those Newtonian principles, as Newton conceived them in the mid-1680s, involved a quite radical shift from the "receiv'd" view—and from what had previously been the view of Newton himself; yet I also believe that Huygens had come closer to them than has generally been recognized.

One fundamental point in respect to which this is so, in my opinion, is the notion of *mass*. No commentator I have encountered considers Huygens to have had a clear conception of mass (as distinguished from weight); and yet just this distinction—between the weight of a body on the one hand, which Huygens does *not* regard as an intrinsic or invariable property of the body, and the quantity that enters into the laws of impact (and of centrifugal force) on the other, which he does so regard—is explicitly affirmed, and argued for, by Huygens in several places, of which the earliest I know dates from 1668.[30] Thus Huygens' law of the conservation of momentum is indeed precisely that of Newton (Corollaries III and IV to the laws of motion).[31]

Just as Huygens succeeded in generalizing his earlier result about direct collision of "hard" bodies to (more or less) arbitrary interactions, so he also undertook to generalize the argument underlying his analysis of centrifugal force. There is a short manuscript of 1675 or 1676[32] in which Huygens considers "the force that acts upon a body to move it when it is at rest, or to augment or diminish its speed when it is in motion." For this kind of force he introduces the term *incitation*; and he argues that incitations deriving from diverse causes—, "weight, elasticity [or 'spring': *ressort*], wind, magnetic attraction"— can be compared quantitatively ("may be equal, the one to the other"). He makes quite clear how this quantitative comparison is to be effected: just as in the case of centrifugal force,[33] equality of incitations can be gauged by the fact that the one is just able to balance the other (and *identity* of "the one" and "the other" is identity, in a suitable senses, of the *state* of the causal agent: "same weight, suspended freely"; "same stretch of same spring"; and so on). The piece ends with a problem and a general principle:

(1) The principle (or "hypothesis"): Equal bodies, constrained to move in equal (that is to say, congruent—in general possibly curved) lines by incitations that are equal at corresponding points of the two paths, will describe their paths in equal times.

(2) The problem: Find the times required to traverse equal spaces, (a) for different incitations acting upon equal bodies, (b) for equal incitations acting upon unequal bodies.[35]

There can be no question but that Huygens was in a position to solve this problem with ease—indeed, the argument of his treatise on centrifugal force *does* solve it, *mutatis mutandis*; and I have no doubt that when Huygens wrote this interrupted manuscript note he knew the solution. But he did not bother to finish the note, just as he did not bother to incorporate the law of the conservation of momentum in the polished manuscript (published posthumously in 1703) of his treatise on impact. The implication seems clear: Huygens did not regard these results—even his "loy admirable de la Nature"—as of an importance comparable with the laws of impact (including that of the conservation of *vis viva*[36] in perfectly elastic impact) or the laws of centrifugal force. And the reason seems clear: the laws of impact *are* the fundamental laws of the communication of motion; the laws of centrifugal force are relevant to all processes that involve *vortex* motion—central, on the one hand, to Descartes's cosmology, and, on the other, to the theory of *gravitation* that Huygens developed (in opposition to Descartes) as early as 1668.[37] The results that come so near to the Newtonian concept of force and laws of motion are, by contrast, of subordinate interest to Huygens. For the task of physical investigation, according to him, is that of studying phenomena with a view to their "mechanical explanation": that is, the development of "hypotheses by motion"[38]—*models*, as we would say, in which all interaction is by pressure or impact—that account adequately for phenomena; which hypotheses are to be treated by testing their observable consequences.[39] In his *Treatise on Light*, Huygens refers to the principles of this program—that is to say, of the "corpuscularian" philosophy—as "the principles accepted in the Philosophy of the present day"; and, a little later, as "the true Philosophy, in which one conceives the causes of all natural effects in terms of mechanical motions"—adding "which in my opinion it is necessary to do, or else renounce all hope of ever understanding anything in Physics."[40]

Let us turn now to the incomparable Newton. At the beginning of his career, in the mid-1660s, Newton seems to have shared Huygens' view of the nature of fundamental explanation: "the lawes of motion" are the laws of impact,[41] and indeed impact of "absolutely hard" bodies;[42] as to *force*, it is "the pressure or crouding of one body upon another."[43] On the other hand, Newton held, already in this early period, a very different philosophy of physical inquiry from that of Huygens. This is exemplified in the optical investigations, in which—to the bafflement of many of his readers, most notably Hooke and Huygens—he claimed (and justly) to have proved with what may be called "experimental certainty,"[44] and quite independently of any mechanical explanation of the nature of light, a set of propositions *about* that nature of far-reaching importance, and of a quite unprecedented character. The essential content of these propositions may be summarized as follows:

(1) *"Light* consists of Rays differently refrangible."[45] Here by "ray" is meant a "part" of light, propagated in a determinate *line.* A beam of light whose rays are of equal refrangibility Newton calls "homogeneal, similar or uniform."[46] The existence of (a continuous range of) such "kinds" of light, and their distinction from ordinary light (which is "heterogeneal"), was established by Newton in one crucial experiment.[47]

(2) "Homogeneal" light possesses a number of intrinsic—or, as Newton says, *"Original* and *connate"*—properties: properties, that is, which cannot be altered by any action exerted upon the light. Among these unmodifiable properties are the refrangibility itself, by which the homogeneous kinds were first distinguished; the "disposition to exhibit this or that particular colour";[48] and—although Newton did not fully discuss this point for some years after his first publication—a particular magnitude, of the nature of a *length* (what we know as the wavelength).[49]

(3) The colorific effect—the visual appearance, or "species" (as Newton calls it)—of light that comes to the eye, is determined by its physical constitution: that is, by the proportions in which it is constituted of the several kinds of rays; there is thus a *mapping* from the set of all possible physical constitutions to the set of perceptible species. But this mapping is by no means one-to-one: many different composite lights cause the same optical perception; and, indeed, Newton says, colors that are "the same in *Specie* with"—that is, indistinguishable by the eye from—those produced by homogeneous light, "may be also produced by composition."[50]

Now I want to call attention to one small point of terminology. Having asserted that although the colors associated with the kinds of homogeneous light are immutable, "seeming transmutations of Colours may be made," because "in . . . mixtures, the component colours appear not," but are "latent,"[51] Newton concludes: "there are therefore two sorts of colours. The one original and simple, the other compounded of these. The Original or primary colours are, *Red, Yellow, Green, Blew,* and a *Violet-purple,* together with Orange, Indico, and an indefinite variety of Intermediate gradations."[52]

The association of the terms *primary, original,* and *simple* had a clear connotation for Newton's audience: *primary* and *original* are approximate philosophical synonyms, denoting a fundamental cause or *principle* (Latin *principium,* rendering Greek ἀρχή, literally "beginning" or "origin"); *simple* denotes a *primary constituent* or "element." All three of these terms, of course, play a most prominent role in Locke's *Essay.* Here is a central passage:

> Qualities . . . in Bodies are, First such as are utterly inseparable from the Body, in what estate soever it be; such as in all the alterations and changes it suffers, all the force can be used upon it, it constantly keeps; and such as Sense constantly finds in every particle of Matter, which has bulk enough to be perceived, and the Mind finds inseparable from every particle of Matter, though less than to make it self singly be perceived by our Senses . . . These I call *original* or *primary* Qualities of Body, which I think we may observe to produce simple *Ideas* in us, viz. Solidity, Extension, Figure, Motion, or Rest, and Number.[53]

I do not at all mean to suggest that Locke has borrowed these terms from Newton. It is now generally supposed, on quite plausible evidence, that Locke's usage in distinguishing qualities into "primary" and "secondary" follows that of Boyle; but my point is that the words belonged to a well-understood philosophical vocabulary, common to the tradition—that is, to *all* the philosophical traditions—from before the time of Aristotle. That this core meaning (that of the "fundamental" and "fundamentally causal")—rather than the notorious distinction between ideas that are "resemblances" of the qualities in things and those that "have no resemblance" of the qualities or "powers" that produce them[54]—is what was central for Locke, appears not only from the passage I have quoted (in which the notion of primary qualities is first introduced), but still more plainly from one of the early drafts of the *Essay,*[55] where no issue of "resemblance" is mooted at all, and where Locke suggests that the only "primary ideas belonging *originally* to bodies"[56] are "extension and cohesion of parts"— giving as his reason that "all the other qualities we observe in, or ideas we receive from, body . . . are *probably* [my emphasis] but the results and modifications of these."

Having perhaps labored this point a bit, let me dwell yet a little further on Locke and primary qualities, to offer an interpretation of that perplexing claim about "resemblance," which provided Berkeley with one of the three major openings for his assault on Locke's philosophy (the other two being Locke's discussions of "abstract" or "general" ideas and of "material substance"). Berkeley's criticism can be paraphrased (or parodied) thus: Locke says that colors, for example, are not in bodies as we perceive them (the colors), but that shapes, for example, are; is he trying to tell us that in point of color bodies do not, but in point of shape they do, *really look* the way they look to us?—That of course would be arrant nonsense. To be sure, philosophers have all too often been guilty of just such nonsense, and I am far from certain that Locke is entirely innocent of it. But I want to suggest that at least a part of what Locke meant by his claim is this (which is not nonsense): that our "ideas" of the primary qualities fit together in a nexus of relations—a structural nexus—that represents "adequately," "faithfully," or "isomorphically" a corresponding nexus in things, at all levels of analysis—most particularly on the fundamental level as conceived in the corpuscularian philosophy, that of the ultimate constituent corpuscles. This, I think, is not a far-fetched, it is even a quite natural rendering (to be sure into a somewhat non-Lockean vocabulary), of his proposition: "That the *Ideas of primary Qualities* of Bodies, *are Resemblances* of them, and their Patterns do really exist in the bodies themselves."[57] On this reading, "resemblance of a pattern" will mean *correspondence to it, point for point* (and relation for relation).

But Locke's claims for primary qualities extend farther than this: not only does he, on the positive side, assert that the ideas of those qualities correctly represent a corresponding metaphysical reality; he also maintains—although (oddly enough, since it is a central tenet of his epistemology) merely as "probable"—that the *only* representations we can form that correspond to a metaphysical reality are those constructed out of ideas of primary qualities. Or rather, Locke throughout the *Essay* in general maintains this; but there are two minute exceptional passages—the results of revisions made, one in the second, one in the fourth edition. I shall have something to say about these passages presently.

There are three points of contrast between Newton and Locke that I want here to call attention to (with two more to come later): First, the qualities Locke calls *primary* and *original* constitute a fixed list; it is clear that Newton's usage is more flexible, since the qualities he attributes to rays of light as primary and original are ones that have been discovered by his experimental investigation itself (and therefore it is reasonable to suppose that further investigations may discover new "primary qualities"). Second, the primary qualities of Locke correspond, in principle, to a subset of "the simple *Ideas* we receive from Sensation and Reflection"—themselves "the Boundaries of our Thoughts; beyond which, the Mind, whatever efforts it would make, is not able to advance one jot; nor can it make any discoveries, when it would prie into the Nature and hidden causes of those *Ideas.*"[58] The properties called "primary" by Newton are, on the contrary, of the class that Locke calls "secondary Qualities, mediately perceivable"[59]—namely, "powers" that manifest themselves only in the perceptible consequences of *physical interaction*—which are not apprehended in any purely passive perception at all.

It might have been suspected that the first of these contrasts was merely terminological, and showed no more than that Newton and Locke differed in their use of the word *primary*. The second contrast, however, makes it clear that a great deal more is at issue. Further insight into what this "more" is is provided by the third contrast, which will require a little more exposition.

In Newton's letter informing the Royal Society of his first optical investigations there occurs, at the point of transition from the account of his discovery of the differences of refrangibility to that of the theory of "the *Origin of Colours*," the cryptic remark that although "a naturalist would scearce expect to see ye science of those [namely, colors] become mathematicall," Newton yet "dare[s] affirm that there is as much certainty in it"—that is, in his mathematical theory of the origin of colors—"as in any other part of Opticks."[60] To this passage Hooke demurred, replying that he could not agree that Newton's theory was "soe certain as mathematicall Demonstrations."[61] In reply, under the heading "That the Science of Colours is most properly a Mathematicall Science," Newton offered a fundamental clarification.[62] His assertion, he says, contained two parts: that his science of colors is mathematical; and that it is as certain as any other part of optics. But this is not to say that this science—any more than the rest of optics—possesses *mathematical certainty*. A science is mathematical, he says, if *from its principles* "a Mathematician may determin all the Phaenomena" it is concerned with; but the "absolute certainty" of any science "cannot exceed the certainty of its Principles"; and "who knows not that Optiques and many other Mathematicall Sciences depend . . . on Physical Principles"—that is, principles "the evidence [for] which . . . is from *Experiments.*" (It will be recalled that in the preface to the *Principia* Newton says the same about geometry itself: that its principles are "fetched from without"; that "Geometry is founded in mechanical practice." I am not aware of any other mathematician or philosopher of the seventeenth century who expressed such a view.[63] Locke quite certainly did not; and neither, more than half a century later, did that arch-empiricist Hume.)

Now, Locke's doctrine about *knowledge*—his "official" doctrine, let me say—is this, that there are "three degrees of Knowledge, *viz.*,

Intuitive, Demonstrative, and Sensitive";[64] of which the first two alone, which "we must search and find only in our Minds," can be "universal," or "general",[65] whereas sensitive knowledge reaches "no farther than the Existence of Things *actually present to our Senses.*"[66] In accordance with this theory of knowledge in the strict sense of the word, Locke repeatedly—but, as it happens, not quite uniformly—denies the very possibility of a *scientific natural philosophy.*[67] The exceptions are all passages in which Locke refers explicitly to Newton. In the *Essay* we find, perhaps ambiguously: "Mr. *Newton*, in his never enough to be admired Book, has demonstrated several Propositions, which are so many new Truths, before unknown to the World, and are farther Advances in Mathematical Knowledge."[68] The possible ambiguity is suggested by the last clause: Locke may refer only to what in the *Principia* Newton calls "mathematical," as contrasted with "philosophical," principles. I am inclined to doubt that he does intend that distinction, because Locke's own interest in the *Principia* certainly did not derive from its purely mathematical content. In any case, the other passages in question admit of no such doubt. In his "Thoughts concerning Education," published in the same year as the *Essay*, one does indeed find the statement: "Natural philosphy, as a speculative science, I imagine, we have none; and perhaps I may think I have reason to say, we never shall be able to make a science of it."[69] But some four pages later Locke qualifies this:

> Though the systems of physics that I have met with afford little encouragement to look for certainty, or science, in any treatise, which shall pretend to give us a body of natural philosophy from the first principles of bodies in general; yet the incomparable Mr. Newton has shown, how far mathematics, applied to some parts of nature, may, upon principles that matter of fact justify, carry us in the knowledge of some, as I may so call them, particular provinces of the incomprehensible universe. And if others could give us so good and clear an account of other parts of nature, as he has of this our planetary world, and the most considerable phaenomena observable in it, in his admirable book "Philosophiae naturalis Principia Mathematica," we might in time hope to be furnished with more *true and certain knowledge* in several parts of this stupendous machine, than hitherto we could have expected.[70]

A second quite definite statement occurs in the public correspondence with Stillingfleet, and bears upon one of the two revisions to Locke's *Essay* that I mentioned earlier. The change is to Book II, chapter 8, §11, which in the first three editions asserted it as "manifest" that "*Bodies operate* one upon another . . . *by impulse,* and nothing else. It being impossible to conceive, that Body should operate on what it does not touch, . . . or when it does touch, operate any other way than by Motion"[71]—thus succinctly expressing the basic tenet of corpuscularianism—and to the first sentence of §12, which began, "If then Bodies cannot operate at a distance . . ." In Locke's letter to Stillingfleet he says, referring to this passage:

> It is true, I say, "that bodies operate by impulse, and nothing else." And so I thought when I writ it, and can yet conceive no other way of their operation. But I am since convinced by the judicious Mr. Newton's incomparable book, that it is too bold a presumption to limit God's power, in this point, by my narrow conceptions. The gravitation of matter towards matter, by ways inconceivable to me, is not only a demonstration that God can, if he pleases, put into bodies powers and ways of operation, above what can be derived from our idea of body, or can be explained by what we know of matter, *but also an unquestionable and every where visible instance, that he has done so.* And therefore in the next edition I shall take care to have that passage rectified.[72]

The emendation Locke actually introduced in his fourth edition is characteristically cautious—to the point, indeed, of *hiding* the issue. In §11 he stills says that "the only way which we can conceive bodies operate in" is "by impulse"; he concludes that this is, manifestly, "how *Bodies* produce *Ideas* in us"; only he makes no statement, now, about the operation of bodies *upon one another*—and, similarly, he drops from §12 the clause about bodies not operating at a distance. I think there is no explicit reference to gravitational attraction in the *Essay.*

There is, however, such a reference in a work originally intended to form a new chapter of the *Essay*: the posthumously published treatise *The Conduct of the Understanding*; and this passage may, I think, fairly be characterized as astonishing. It occurs in section 43, whose title is "Fundamental Verities"; and one might think that a very famous piece of eloquence in Kant's *Critique of Practical Reason* was derived from it. "There are," Locke tells us, "fundamental truths that lie at the bottom, the basis upon which a great many others rest, and in which they have their consistency. These are teeming truths, rich in store, with which they furnish the mind, and, like the lights of heaven, are not only beautiful and entertaining in themselves, but give light and evidence to other things that without them could not be seen or known." He gives just two examples of such "fundamental," "teeming" (that is, pregnant) truths. The first is "that admirable discovery of Mr. Newton, that all bodies gravitate to one another, which may be counted as the basis of natural philosophy; which of what use it is to the understanding of the great frame of our solar system, he has to the astonishment of the learned world shown, and how much further it would guide us in other things, if rightly pursued, is not yet known." And what is the second fundamental truth, deemed worthy by Locke of comparison with Newton's admirable discovery? It is "our Saviours great rule, that *we should love our neighbor as ourselves*": —this "is such a fundamental truth for the regulating of human society, that I think by that alone one might without difficulty determine all the cases and doubts in social morality." (It should be remembered that in the *Essay* Locke has affirmed his belief that morals—like mathematics, and *unlike* natural philosophy—could be made a demonstrative science.)

The third contrast I have drawn, then, is between two different conceptions of what is required for a systematic science. Locke's curious ambivalence on the point—why does he not, although admiring Newton's discoveries, refuse to allow them the name of "science" in its strict sense, on the grounds that they do not rest on principles evident to intuition?—suggests that he is less firmly committed to his "official" epistemology (and metaphysics) than most of his commentators take him to be. For instance, Michael Ayers, in a very acute and stimulating article on Locke,[73] has declared that "if he let in the possibility that powers or phenomenal properties should belong to things as a matter of brute or miraculous fact not naturally intelligible, Locke's whole carefully constructed philosophy of science and his support for the corpuscularian case against the Aristotelians would collapse."[74] But this possibility—indeed the certainty that it is true, in the case of the power of gravitational attraction—is exactly what, in the passage quoted above, Locke tells Stillingfleet that Newton has convinced him of.

The two issues—whether "primary" qualities, or at least our *knowledge* of primary qualities, need be directly and simply related to our modes of perception; and whether science need be grounded in what is immediately perceived ("intuited") by the mind in contemplating its "ideas"—are very closely connected. It has been characteristic of an influential branch of modern empiricism to adopt the affirmative position on each of these issues; and thus to deny a central distinction of that great ancient empiricist Aristotle: the distinction between what is "first, and better known, in nature" and what is first, and better known, "to us." It is possible that a recognition of such a distinction is implied by one cryptic (and to me puzzling) remark by Locke in the *Essay*: "'Tis fit to observe," he says, "that Certainty is twofold; *Cer-*

tainty of Truth, and *Certainty of Knowledge. Certainty of Truth* is, when Words are so put together in Propositions, as exactly to express the agreement or disagreement of the *Ideas* they stand for, as really it is."[75] What is the difference Locke implies here between what he calls "certainty of truth"—which does *not* imply certain *knowledge*—and just "truth"? His emphasis seems to be on the phrase "as it really is"; and this may, although it surely is obscure, be a remote echo of Aristotle's "better known *in nature.*"[76] However this may be, it is certainly characteristic of what I have called Locke's "official" doctrine to conflate—so far as humanly possible science is concerned—epistemological priority and metaphysical priority, and to find in the evident inadequacy of this correspondence an impassable boundary to human knowledge.

On this issue, then, Newton may be said to stand with Aristotle. Not, however, on another—(and closely related)—one; for Aristotle and Locke—and Descartes, and, to a degree, Huygens—agree on this point: that genuine science is possible only on the basis of genuinely, "absolutely," first principles (of which Aristotle says—and indeed of "causes" in general, whether "first" or "intermediate"—that there must be *in toto* a *finite* and *known system* of causes if there is to be science at all).[77] Descartes thought he had definitively established such a system through an analysis of qualitative experience in the light of principles innate to the mind. Huygens came to reject most of Descartes's analysis, and to regard fundamental physical knowledge as something to seek on probable, rather than certain, evidence; but, as we have seen, he continued to hold firmly to the view that the whole possibility of progress in physics depended upon the reduction of all phenomena to "mechanical" interaction "by motion." Huygens offers us no argument on this point; Locke, the mere assertion that this is the only mode of action that the "idea of body" makes intelligible. Newton, however, no more believes in the intrinsic and unique "intelligibility" of this mode of interaction than does Hume.[78] Newton writes:

> We no otherwise know the extension of bodies than by our senses, nor do these reach it in all bodies; but because we perceive extension in all that are sensible, therefore we ascribe it to all others also. That abundance of bodies are hard we learn from experience. And because the hardness of the whole arises from the hardness of the parts, we therefore justly infer the hardness of the undivided particles not only of the bodies we feel but of all others. That all bodies are impenetrable, we gather not from reason, but from sensation . . . That all bodies are moveable, and

endow'd with certain powers (which we call the *vires inertiae*) of persevering in their motion or in their rest, we only infer from the like properties observ'd in the bodies which we have seen. The extension, hardness, impenetrability, mobility, and *vires inertiae* of the whole, result from the extension, hardness, impenetrability, mobility, and *vires inertiae* of the parts: and thence we conclude the least particles of all bodies to be also extended, and hard, and impenetrable, and moveable, and endow'd with their proper *vires inertiae*. And this is the foundation of all philosophy.[79]

The "foundation of all philosophy" (by which he means "of all physics")—that part, we may say, of metaphysics, that is required for physics—is itself founded on, or derived from, not rational insight, but empirical evidence. The fourth of the contrasts I wish to draw between Locke and Newton concerns this metaphysics itself (rather than its epistemological grounds). It is related to the second revision to Locke's *Essay* known to have been made under the influence of Newton.

In the first edition of the *Essay*, in attacking the argument that matter must be eternal because its creation *ex nihilo* is inconceivable, Locke had countered that the creation of a mind or "spirit" is as much beyond our comprehension as that of a body.[80] In the second edition this was revised to say that in fact, "if we would emancipate ourselves from vulgar Notions," we might be able to form "some dim and seeming conception how Matter might at first be made . . . : But to give beginning and being to a Spirit, would be found a more inconceivable effect of omnipotent Power." He declines, however, to particularize, saying that to do so "would perhaps lead us too far from the Notions, on which the Philosophy now in the World is built."

We know from Locke's French translator, Pierre Coste, that the intimated radical departure from the received philosophy was sketched to Locke, in conversation, by Newton;[81] and we now have, in the fragmentary manuscript "De gravitatione et aequipondio fluidorum et solidorum in fluidis," published by the Halls[82] (together with, unfortunately, a most defective translation), Newton's own exposition of this heterodox metaphysics. I have discussed this elsewhere in more detail;[83] here let me only remark, first, that Newton's explicit aim, in his discussion of "corporeal nature" in that text, is to free the conception of body from two scholastic notions he regards as unintelligible: that of a *formless substrate*, and that of a "substantial form" inhering in such a substrate (or "unintelligible substance") as subject; second, that what he offers instead of the "unintelligible substance" and its unintelligible "substantial form" is the conception of *spatial distributions of clearly conceived attributes*— space or extension, of which (Newton says) we have "an exceptionally clear Idea," playing the role of "subject," and the distributed attributes that of "form"; third, that it is a crucial part of this view of bodies as constituted by "spatially distributed attributes"—what we should now call "fields" on space—that these have, as part of their formal constitution, definite "laws of motion," or of propagation from one region of space to another; fourth, corresponding to Locke's remark that the creation of a mind is *more inconceivable* than that of a body, that Newton, although he suggests the possibility of an analogous conception of God himself, free of any unintelligible notion of a "substantial subject" in which his attributes inhere, quickly adds that the defect of our ideas of God's attributes and even of our own mental powers makes it "rash to say what may be the substantial basis of minds."

From the "official" point of view of Locke's *Essay*, the creation of a substance exceeds human comprehension precisely because, in that view, the "idea" of such a substance inescapably includes what Locke himself characterizes as the to us necessarily obscure and confused "idea of substance in general."[84] The "Philosophy now in the World"—the corpuscularian philosophy—while rejecting "prime matter" and "substantial forms" (or "occult qualities"), still required that obscure idea of substance. And this Newton's more radical metaphysics contrives to do without.

It is of some importance to note that the metaphysics of "De gravitatione" attributes to bodies, alongside Locke's "primary and original" attributes of extension, solidity (or impenetrability), and mobility,[85] two others, not on Locke's list: namely, mass or *vis inertiae*, and the power of stimulating perceptions in a mind. To be sure, mass is not there mentioned explicitly. It is, however, implied by the requirement that there be definite laws of motion of the impenetrable regions of space; and this may serve to remind us that Locke's own corpuscularian theory of the "intelligible" interaction "by impulse" presupposes this attribute of mass—which cannot be construed to correspond to a simple idea, but can only be understood as a power "mediately perceivable." In this sense, even apart from the questionable "idea of substance," the Lockean view defended by Ayers has to be regarded as not fully coherent.

The account of the creation of matter in "De gravitatione" is parallel to the much better known passage on the same subject in Newton's *Opticks*.[86] The latter, with its reference to "solid, massy, hard, impenetrable, moveable Particles," is reticent about the deeper meta-physical foundations; but the correspondence is made evident by a cryptic phrase which, I suggest, becomes clear when the two accounts are juxtaposed. Newton says that the "principles" he has spoken of are to be regarded, "not as occult Qualities, supposed to result from the specific Forms of Things, but as general Laws of Nature, *by which the Things themselves are form'd*" (my emphasis). What does this last phrase mean? It means precisely that Newton proposes to replace, as constituting the essence of corporeal things, the "occult qualities" and "substantial forms" of the scholastics with *clear* forms, specifiable as "laws of nature" whose truth is evinced for us by phenomena. (There is actually a little more that needs to be said for a careful analysis of Newton's statement, but I have not time for that here; the main qualification will be implied by my concluding remarks—and last contrast with Locke.)

In one positive point the passage in the *Opticks* differs significantly from that in "De gravitatione." The "principles," or constitutive laws of nature, that Newton most especially means to defend against the charge that they are "occult" are, not those of solidity, impenetrability, rigidity, mobility, and inertia, with the "passive Laws of Motion" that characterize the *vis inertiae,* but a new set, not at all appearing in "De gravitatione": "certain active Principles, such as is that of Gravity, and that which causes Fermentation, and the Cohesion of Bodies." This very crucial amendment is the result of the great investigation that gave us the *Principia.*

That investigation may be succinctly described as having three phases. In the first phase—already partially accomplished in the famous plague years of 1665–66—Newton found, on the basis of Kepler's so-called third law and the approximation of planetary motion as uniform and circular, that (to use Huygens' word) the "incitations" of the planets toward the sun are inversely as the squares of the distances; and found also, transferring this law to the moon, that the latter's "incitation" toward the earth can be identified with its *weight* toward the earth. With this result, and with the further inference that (a) the "incitation" of the planets also is to be identified with their *weight toward the sun,* and that (b) weight in general obeys this law of the inverse square, Huygens declared himself in full agreement; intimated, not unjustly, that he himself could have made this discovery if he had had the boldness to consider the possibility of weight acting at such vast distances; and expressed his great admiration of Newton for having done so.[87] But Newton's move in what I am calling the second phase is far bolder, and is the turning point of the investigation: Quite setting aside the standard view that the fundamental laws of motion apply to the fundamental interactions "by impulse," Newton treats the gravitation of a body A toward a body B as a case of *interaction* between A and B, and applies directly to it his own version of Huygens' "wonderful law of Nature," the conservation of momentum—namely, Newton's third law of motion. From this, with some qualitative considerations regarding weight, he infers with breathtaking swiftness the existence of a *universal* law of gravitation between all pairs of corporeal particles in the universe.[88]

The third and final phase of the investigation takes up the major part of Book III of the *Principia*—and the major part of astronomy for the next two-hundred-odd years. It consists in the deduction of the detailed consequences of Newton's new law, and their detailed comparison with increasingly precise data. It is, as I have remarked elsewhere,[89] on the success of this third phase that Newton rests his case for the correctness of his result: the critical second phase of the investigation must be regarded as *heuristic,* not as *demonstrative.*

But the third phase—or its success—had in turn a heuristic consequence of profound importance: it led Newton to the amendment of his metaphysics that I have already cited, and therewith to a new program for physics. This is adumbrated in the introductory sections of the *Principia*, preceding Book I: the "Definitions" and "Laws of Motion"; and is summarized in Newton's preface to the work. Its central notion is that of a *vis naturae* or *potentia naturalis*—a "force of nature" or "natural power." The three laws of motion (as Newton makes explicit in the *Opticks*) constitute, together, the characterization of one of these forces: the *"Vis inertiae,"* which is "a passive Principle."[90] The remaining forces, the "active Principles," are what, according to this program, it is the chief task of natural philosophy to seek to discover. And it is within this scheme that the deep importance of Huygens' "incitation," which Newton calls "motive force," really emerges: for the *action upon a body* of a force of nature—what Newton calls an *impressed* force—has this "incitation," or "motive quantity," as its appropriate measure; the latter therefore enters into the expression of every fundamental law of nature.

The account I have just sketched illustrates the last of the contrasts I wish to make with Locke, and with the received corpuscularian view generally. Perhaps it is already implied by the other four contrasts. Artistotle, as I have said, thought that the very possibility of science depends upon the possession of first principles; and on this point, as I have also said, Locke and Huygens agree with him; but Newton does not. In Aristotle's view, first principles themselves are discovered in a process that *precedes* science—a process he describes as "dialectical" rather than "scientific." In these terms (which are foreign to Newton's own usage), Newton may be said to agree rather with Plato, for whom science itself was "dialectical." The double-facedness of the result in the *Principia*—"forward" to the greater mastery of phenomena, "backward" to new principles—is a perfect illustration. But that it is a genuine illustration of a philosophy of inquiry that reigned over Newton's entire creative career is apparent from his writings, both early and late. In his inaugural lectures as Lucasian Professor at Cambridge in 1670 (when he was not yet twenty-eight), Newton declares, concerning his "mathematical science of colors": "I hope to show—as it were, by my example—how valuable Mathematics is in natural Philosophy. I therefore urge Geometers to investigate Nature more rigorously, and those devoted to natural science to learn Geometry first. Hence . . . truly with the help of philosophizing Geometers and geometrizing Philosophers, instead of the conjectures and probabilities that are hawked on all sides, we shall at last achieve a natural science supported by the highest evidence."[91] In the final query of the *Opticks*, to which I have already referred, and which first appeared in the Latin edition of 1706 (when Newton was approaching sixty-four), after proposing the scheme of passive and active forces (or laws), Newton conspicuously refrains from claiming for these any *final*, any *ultimately "foundational"* status: "To tell us," he says,

> that every species of Things is endow'd with an occult specifick Quality by which it acts and produces manifest Effects, is to tell us nothing: But to derive two or three general Principles of Motion from Phaenomena, and afterwards to tell us how the Properties and Actions of all corporeal Things follow from those manifest Principles, would be a very great step in Philosophy, though the Causes of those Principles were not yet discover'd. And therefore I scruple not to propose the Principles of Motion above-mention'd, they being of very general Extent, and leave their Causes to be found out.[92]

Perhaps the simplest of all Newton's statements of what I have called his dialectical conception of science is that in the preface to the *Principia*. As in the other passages I have cited, here too what Newton offers is not a proposed *foundation* for physics, but a *framework within which physical investigation may be possible*. But the subtlest dialectical turn is Newton's intimation—which has been dramatically confirmed in our own century—that such investigation may lead, not only to new laws and deeper causes, but to a revision of the framework itself. After setting forth his new program for natural philosophy, Newton concludes—in words that I have quoted on more than one previous occasion, and which I have always found moving—"But I hope the principles here laid down will afford some light either to that, or some truer, method of Philosophy."

Notes

1. Christiaan Huygens, *Oeuvres complètes de Christiaan Huygens*, Société hollandaise des Sciences, vol. I (La Haye: Martinus Nijhoff, 1888), p. 125; the quoted stanza also appears ibid., vol. XVI (1929), p. 4.
2. Ibid., vol. X (1905), pp. 402–406.
3. J. L. Heilbron, *Electricity in the 17th and 18th Centuries* (Berkeley: University of California Press, 1979), p. 31.
4. René Descartes, *Principia Philosophiae*, pt. II, ¶64.
5. Ibid., pt. I, ¶53; pt. II, ¶4.
6. Ibid., pt. II, ¶45.
7. Ibid., ¶23.
8. Ibid., ¶¶46–52. The first rule is correct for the elastic impact of equal bodies colliding "symmetrically"; the third, the fourth, and the first part of the seventh are correct for perfectly inelastic impacts of the three sorts they treat; the remaining ones are never true. And the relationships among these rules for the several cases are quite inconsistent with the theory of the nature of motion itself professed by Descartes in ¶¶25–30; for instance, the case in which a smaller body in motion strikes a larger one at rest (the fourth rule, ¶49), and that in which a larger body in motion strikes a smaller one at rest (the fifth rule, ¶50), are treated as fundamentally different, although ¶29 has told us that "we cannot understand a body AB to be transferred from the vicinity of a body CD, without at the same time also understanding the body CD to be transferred from the vicinity of the body AB: and that exactly the same force and action is required for the one as for the other"—in other words, that to which of two bodies motion is ascribed, to which rest, is entirely conventional, and corresponds to no physical difference. (Yet again, this stands in contradiction to ¶44, which asserts that a relation of "contrariety" holds, not between one motion and another, but between motion and rest.) It may incidentally be noted that Descartes's seven rules do not cover all possible cases of impact: he fails to consider bodies of unequal size moving toward one another with unequal speeds.
9. Ibid., ¶53. Descartes does not say explicitly that his rules apply *in vacuo*; but he does imply that they hold only for colliding bodies that are "separated from all others" (cf. n. 10 below). This cannot mean merely that the bodies do not *cohere* with those that surround them; for according to Descartes's theory of cohesion (ibid.), this in turn would mean that the bodies surrounding the given ones were moved *diversely* from the latter (and necessarily then—if the plenum is to be maintained—diversely from one another); which (again by Descartes's theory) is to say that the ambient medium is *fluid*; and it is to the effect of just such an ambient medium that Descartes attributes the *failure* of his rules.
10. Ibid., end of ¶52 in the French version: "And the demonstrations . . . are so certain that even if experience were to appear to show us the opposite, we should nevertheless be obliged to place more trust in our reason than in our senses." (This passage, added in the French translation by the Abbé Picot, is ascribed to Descartes himself, on the basis of a letter to Mersenne of April 20, 1646, indicating that he was engaged in "clarifying my laws of motion" for Picot's translation.) A reason for

the divergence of the actual behavior of colliding bodies from his rules of impact quite different from that given in the *Principia*, and perhaps even more astounding, is given by Descartes in a letter of February 17, 1645, to Claude Clerselier. Whereas in the passage cited in n. 9 above Descartes attributes this divergence to the circumstance that "there cannot be any bodies in the world that are separated from all others," in the letter to Clerselier he says almost the opposite: "In those rules, by a body devoid of motion, I intend a body that is not at all in the act of separating its surface from those of the other bodies that surround it, and, in consequence, that forms a part of another, larger, solid body." How this picture of the body at rest as embedded in a *solid ambient medium* is to be reconciled, e.g., with the discussion in the fifth rule of a larger body striking a *smaller* one at rest—or with the stipulation in the fourth rule that the body at rest may be taken to be just *slightly* larger than the one striking it—is not further explained.

11. Descartes, letter of February 26, 1649, to Pierre Chanut.
12. Letter of August 22, 1634. (The addressee—not indicated in the extant source— was tentatively identified by Adam and Tannery as Isaac Beeckman; but in their supplement to the correspondence—volume 10 of their edition of the *Oeuvres* of Descartes—they express substantial doubt on this point.)
13. Huygens, *Oeuvres complètes*, XVI, 30/31 (French translation and Latin original, respectively), Hypothesis I.
14. Ibid., pp. 32/33, Hypothesis III. Huygens' careful formulation is worth quoting: "The motion of bodies, and equal or unequal speeds, are to be understood as relative, having relation to other bodies that are considered as resting—although both the latter and the former may be involved in some other common motion. And therefore when two bodies collide, although both together may be subject to some further equable motion, they do not drive one another any differently, in relation to one who is carried by the same common motion, than if that additional motion were entirely absent." It is especially noteworthy that although the first sentence seems to ground this principle in the philosophical view—which Huygens certainly held—that the only *intelligible concept* of motion is that of one body relative to another, his insight—and conscience—as a physicist compelled him to restrict the principle to "additional common motions" that are *equable:* in the manuscript from which the treatise on impact was printed, which was written out by an amanuensis, the word *aequabili* is inserted in Huygens' own hand (see ibid., p. 33, n. 3).
15. Ibid., pp. 56, ll. 3ff./57, ll. 14ff.
16. Cf. ibid., pp. 164–165, n. 2. For the converse, see ibid. XVIII (1934), 250, ll. 9ff./ 251, ll. 8ff. (with reference back to Hypothesis I, pp. 246/247).
17. See the two manuscript pieces, ibid., XVI, 161–167; and cf. p. 181, first paragraph.
18. Ibid., p. 181.
19. That is, for one moving "equably" (see n. 14).
20. Huygens, in his argument, specifies a particular choice of v that will violate the inequality when the quantity in question is nonzero.
21. This relation appears in the major (posthumously published) treatise—which deals only with "hard" bodies—as Proposition IV. It is there deduced with the help of another assumption (Hypothesis V). But Huygens had earlier taken Proposition IV as a hypothesis (see ibid., p. 40 n. 2 and p. 42 n. 1).
22. Isaac Newton, *Philosophiae Naturalis Principia Mathematica*, Scholium to the laws of motion and their Corollaries.
23. Huygens, *Oeuvres complètes*, XVI, 254/255ff. My exposition does not follow the same order as that of Huygens: he is not explicit about the train of thought that motivates his argument.

24. Huygens does not mention the *buoyant* effect of the air, which really demands a little more argument.—The editors of this volume of the *Oeuvres complètes* of Huygens annotate Huygens' reference to the experiments of Riccioli by citing the latter's *Almagestum Novum* (1651 ed.), I, 381–397; they give no reference to any published or manuscript record of Huygens' own experiments testing Galileo's law of falling bodies. The nearest approach to such an account that I have found is a series of records published in volume 17 (1932) of the *Oeuvres complètes*, pp. 278–284. Three of these, dating from October and November 1659, describe experiments designed to determine the distance a body falls in a given time; but the only times in fact used are 1/2 second and 3/4 second—the aim being to determine (in effect) the acceleration of free fall (in the form: the distance covered in one second by a body falling freely from rest)—and in the latest of these pieces Huygens expresses his distrust of the method used, and his reliance instead upon the value obtained from the relation of length to period of a conical pendulum. A fourth piece in this series, dated simply to 1659, deals with a qualitative experiment whose aim is to show that the initial velocity of a body falling from rest is zero. The two remaining pieces in the series date from August or September 1664, and describe an apparatus for measuring distances fallen in a given time; but no numerical data are given, and the editors of the volume state (p. 247) that Huygens apparently did not in fact build apparatus of the sort he describes or have such built. In short, none of this material directly illuminates Huygens' reference to his own experimental confirmation of Galileo's law.

25. The string is supposed parallel to the inclined plane. Huygens asserts (loc. cit., pp. 256/257), but does not argue for, the proportionality of "tug" to acceleration. He could have given such an argument, by referring to the magnitude of the weight required for equilibrium if the string passes over a pulley and an equilibrating weight is hung vertically from its other end. Experiments, then, on the relation of acceleration to the inclination of the plane, compared with the relation of equilibrating weight to that inclination, would support the assertion. (There remains the further consideration of the relationship of the acceleration of a body, presumably *rolling* on an incline, to that of a *freely falling* body—or to that "ideally" expected of a body *sliding* without friction on an incline. I am not aware that this is a matter ever addressed by Huygens.)

26. This may have been the occasion of Huygens' first reflection upon the involutes of curves—a notion he introduced publicly, and put to ingenious use, in his book on the pendulum clock.

27. Ibid., pp. 266/267.　　　　　　　　　　　　　　　　　．

28. Huygens, it should be made clear, does *not* enunciate a principle as general as this (but see below, on "incitation"). He does, however, prove (although not in explicit algebraic form) that a body of weight W, moving in a circle of radius r with velocity v, will exert a centrifugal force F such that $F:W = v^2/r:g$, where g is the acceleration of free fall (at the location where the weight of the body amounts to W: for Huygens does not think that weight is independent of location). So for bodies in general in whirling motion, F is proportional to $(W/g)\cdot(v^2/r)$. Furthermore (see discussion below), Huygens *does* believe that W/g is constant—i.e., is an intrinsic property of a body.

29. Newton, *Principia*, Scholium to the laws of motion and their corollaries.

30. Huygens, *Oeuvres complètes*, XIX (1937), 627: "Moy je dis que chasque corps a de la pesanteur suivant la quantité de la matiere qui le compose . . . Cela paroit de l'effet de l'impulsion qui suit exactement la raison de la pesanteur des corps." Here, to be sure, the distinction is *presupposed* rather than argued for. But one finds an argu-

ment for it in a passage of 1669, ibid., pp. 637–638. (Cf. also vol. XXI [1944], pp. 382 and 458.)

31. The conclusion explicitly stated by Huygens in the passage cited in n. 18 above corresponds to Newton's Corollary IV: conservation of motion of the center of gravity.

32. Huygens, *Oeuvres complètes*, XVIII, 496–498.

33. To prevent misunderstanding: the *strategy* is the same as in the case of centrifugal force; Huygens himself does not mention centrifugal force in this manuscript.

34. This is clearly implied by Huygens' examples, although he does not quite make it explicit.

35. My interpretation of the logical connections among Huygens' three fragmentary clauses:

incitations differentes uniformes et corps egaux.
quel temps par des espaces egaux.
incitations egales sur des corps inegaux.

—loc. cit., p. 498.

36. Leibniz's term, of course; not Huygens'.

37. Huygens, *Oeuvres complètes*, XIX, 625ff.

38. See *Philosophical Transactions* [scil., *of the Royal Society of London*] no. 96 (July 21, 1673), 6086; in I. Bernard Cohen, ed., *Isaac Newton's Papers and Letters on Natural Philosophy* (Cambridge, Mass.: Harvard University Press, 1958), p. 136; in H. W. Turnbull, ed., *The Correspondence of Isaac Newton*, vol. I (Cambridge: Cambridge University Press, 1959), 255 (in this last place, quoted by Oldenburg to Newton in Huygens' original French).

39. Huygens, *Treatise on Light*, trans. Silvanus P. Thompson (reprint: Chicago: University of Chicago Press, 1945), pp. vi–vii; *Oeuvres complètes*, XIX, 454–455.

40. Ibid. (Thompson trans.), pp. 2, 3; *Oeuvres complètes*, pp. 459, 461.

41. *Unpublished Scientific Papers of Isaac Newton*, ed. A. Rupert Hall and Marie Boas Hall (Cambridge: Cambridge University Press, 1962), pp. 157ff.; also in John W. Herivel, *The Background to Newton's Principia* (Oxford: Clarendon Press, 1965), pp. 208ff.

42. Loc. cit., Hall and Hall, p. 162; Herivel, p. 213.

43. Herivel, *Background to Newton's Principia*, p. 138.

44. Cf. the concluding paragraph of Newton's reply to Hooke's comments, in Newton, *Correspondence*, I, 187. (This passage is not to be found in the Cohen edition of Newton's papers and letters, cited above, n. 38, which reproduces Newton's reply as published by Oldenburg in the *Philosophical Transactions*: Oldenburg omitted the paragraph in question.)

45. Newton, *Correspondence*, I, 95; *Papers and Letters*, p. 51.

46. *Correspondence*, I, 96, 292; *Papers and Letters*, pp. 53, 140.

47. Newton's claim, which I am prepared to defend.

48. *Correspondence*, I, 97; *Papers and Letters*, p. 53.

49. That Newton was in possession of this result in 1672 (although he did not expound it in detail until December 1675—see *Papers and Letters*, pp. 193–198, 204–206 (Observations 5–7), 207–208 (Observations 14–15), 210 (Observation 16); less completely in *Correspondence*, I, 377–383—is clear from the first paragraph of §3 of his reply to Hooke (*Correspondence*, I, 174–175; in *Papers and Letters*, this is §4, pp. 120–121).

50. *Correspondence*, I, 98 (¶6); *Papers and Letters*, pp. 54–55.

51. *Correspondence*, I, 97–98; *Papers and Letters*, pp. 53–54.

52. *Correspondence*, I, 98; *Papers and Letters*, p. 54.

53. John Locke, *An Essay Concerning Human Understanding*, II, viii, §9; quoted from the edition of Peter H. Nidditch (Oxford: Clarendon Press, 1975, 1979), pp. 134–135.

54. Ibid., II, viii, §15, p. 137.

55. Locke, *An Essay Concerning the Understanding, Knowledge, Opinion, and Assent*, ed. Benjamin Rand (Cambridge, Mass.: Harvard University Press, 1931), now generally known as Draft B of the *Essay*; see pp. 198–199.

56. This is an example of Locke's habit—which he explicitly notes—of using the word *idea* for what, when he is more precise, he calls *quality*.

57. Locke, *Essay*, II, viii, §15, p. 137. The word *pattern*, by the way (see the *Oxford English Dictionary*) is originally the same word as *patron*, hence with the root meaning "father": "original"; the two spellings were not fully differentiated until c. 1700.

58. Locke, *Essay*, II, xxiii, §29, p. 312.

59. Ibid., viii. §26, pp. 142–143.

60. Newton, *Correspondence*, I, 96; the passage was omitted by Oldenburg from the version published in the *Philosophical Transactions* and is therefore not to be found in *Papers and Letters*.

61. Ibid., p. 113.

62. Ibid., pp. 187–188.

63. In particular connection with the theory of parallels, a similar conviction was expressed in a Göttingen dissertation of 1763 by G. S. Klügel. See Roberto Bonola, *Non-Euclidean Geometry: A Critical Study of its Development*, trans. H. S. Carslaw (reprint: New York: Dover Publications, 1955), pp. 50–51 and p. 44, n. 3. Gauss appears to have been the first mathematician of stature (after Newton) to have come—and only after a struggle—to hold seriously the view that the grounds of geometry are empirical.

64. Locke, *Essay*, IV, ii, §14, p. 538.

65. Ibid., iii, §31, p. 562.

66. Ibid., §5, p. 539 (emphasis added).

67. Cf., e.g., ibid., iii, §26, pp. 556–557, and xii, §§9–10, pp. 644–645.

68. Ibid., IV, vii, §11, p. 599.

69. Locke, *Some Thoughts Concerning Education*, §190, in *The Works of John Locke*, new ed., corrected, in 10 vols. (London: Thomas Tegg et al., 1823), IX, 182.

70. Ibid., §194, p. 186 (emphasis added). Note, too, the sentence that follows, which evidently alludes to the assurance given to Locke by a Dutch mathematician—reportedly Huygens—that Newton's mathematical demonstrations were correct. In his biography of Huygens in volume 22 of the *Oeuvres complètes*, J. A. Vollgraff expresses doubt that Huygens and Locke had met during the latter's sojourn in Holland (see p. 744, n. 26); but what other Dutch mathematician would have been competent to offer that assurance?

71. This is the wording of 1st edition; in editions 2–3 there is a slight change, which seems to make the statement a little less clear (see Nidditch edition, pp. 135–136, and critical apparatus to 135, l. 31–136, l. 2).

72. In *The Works of John Locke*, 9 vols., 12th ed. (London: C. and J. Rivington et al., 1824), vol. III, "Mr. Locke's Reply to the Bishop of Worcester's Answer to his second Letter," pp. 467–468 (emphasis added); cf. the edition of Locke's *Essay* by Alexander Campbell Fraser (reprint: New York: Dover Publications, 1959), I, 171, n. 1.

73. M. R. Ayers, "The Ideas of Power and Substance in Locke's Philosophy," *The Philosophical Quarterly* 25 (1975), 1–27.

74. Ibid., p. 22.
75. Locke, *Essay*, IV, vi, §3, p. 579.
76. The suggestion I have made here seems to me at best marginally convincing; nevertheless, I wish to offer a further argument in its favor. Locke's account of truth (ibid., v, §§2, 5–8, pp. 574, 575–578) is essentially contained in these words: "When *Ideas* are so put together, or separated in the Mind, as they, or the Things they stand for do agree, or not, that is, as I may call it, *mental Truth*. But *Truth of Words* is something more, and that is the affirming or denying of Words of one another, as the *Ideas* they stand for agree or disagree: And this again is twofold. Either *purely Verbal*, and trifling, . . . *or Real* and instructive . . . Truth, as well as Knowledge, may well come under the distinction of *Verbal* and *Real*; that being only *verbal Truth*, wherein Terms are joined according to the agreement or disagreement of the *Ideas* they stand for, without regarding whether our *Ideas* are such, as really have, or are capable of having an Existence in Nature. But then it is they contain *real Truth*, when these signs are joined, as our *Ideas* agree; and when our *Ideas* are such, as we know are capable of having an Existence in Nature." This seems clearly enough to warrant the reading of the phrase quoted in the text—"the agreement or disagreement of the *Ideas* [the words stand for] *as it really is*"—as demanding a correspondence of the "Ideas" (which agree or disagree) with some correlate *in rerum natura*. But the trouble is that it fails to suggest why the satisfaction of this demand is appropriately called a kind of *certainty*; especially in view of the fact that the condition Locke states for what he calls *"Certainty of Truth"* is precisely the same as the one he has stated just previously for what he calls *"real Truth."* Perhaps, after all, no weight should be attached to this expression, and it should simply be taken as one instance among many of Locke's almost studied imprecision of usage.
77. Aristotle, *Metaphysics*, II, 2.
78. Cf., e.g., David Hume, *An Enquiry concerning Human Understanding*, §4, pt. 2, in the edition of L. A. Selby-Bigge (3d ed., rev. by P. H. Nidditch) (Oxford: Clarendon Press, 1975), §25, pp. 29–30.
79. Newton, *Principia*, Book III, Rule III (of the Rules of Philosophizing). Quoted from the original (unrevised) translation of 1729 by Andrew Motte (reprint: London: Dawsons of Pall Mall, 1968).
80. Locke, *Essay*, IV, x, §18, p. 628 (with critical apparatus to ll. 30ff.).
81. See Locke, *Essay*, ed. Fraser, II, 321, n. 2.
82. Newton, *Unpublished Scientific Papers*, pp. 90–121 (Latin text): pp. 121–156 (translation).
83. Howard Stein, "On the Notion of Field in Newton, Maxwell, and Beyond," in *Historical and Philosophical Perspectives of Science*, ed. Roger H. Stuewer, Minnesota Studies in the Philosophy of Science, vol. V (Minneapolis: University of Minnesota Press, 1970), pp. 273ff.; "On Space-Time and Ontology," in *Foundations of Space-Time Theories*, ed. John S. Earman, Clark N. Glymour, and John J. Stachel, Minnesota Studies in the Philosophy of Science, vol. VIII (Minneapolis: University of Minnesota Press, 1977), pp. 395ff.; "On Metaphysics and Method in Newton" (unpublished).
84. See, e.g., Locke, *Essay*, II, xii, §6, p. 165: "The *Ideas* of *Substances* are such combinations of simple *Ideas*, as are taken to represent distinct particular things subsisting by themselves; in which the supposed, or confused *Idea* of Substance, such as it is, is always the first and chief"; and cf., in Locke's own index to the work, the following entry under the head "SUBSTANCE" (p. 745): "The confused Idea of S. in general makes always a part of the Essence of the Species of Ss."

85. Ibid., II, xxi, §73, p. 286.
86. Newton, *Opticks* (reprint: New York: Dover Publications, 1952), pp. 401–403. It should be noted that by *solid* Newton does not mean (as Locke does) "impenetrable": in Newton's usage, *solid* means what it ordinarily means to us: "cohesive" (and when, further, Newton wishes to characterize a body as what we call *rigid*, he describes it as [both solid and] *hard*).
87. See Huygens, *Discours de la Cause de la Pesanteur,* in his *Oeuvres complètes,* XXI (1944), p. 472: "Je n'avois point etendu l'action de la pesanteur à de si grandes distances, comme du Soleil aux planetes, ni de la Terre à la Lune; parce que les Tourbillons de Mr. Des Cartes, qui m'avoient autrefois paru fort vraisemblable, & que j'avois encore dans l'esprit, venoient à la traverse. Je n'avois pas pensé non plus à cette diminution reglée de la pesanteur, sçavoir qu'elle estoit en raison reciproque des quarrez des distances du centre: qui est une nouvelle & fort remarquable proprieté de la pesanteur, dont il vaut bien la peine de chercher la raison." And, after having discussed the objections raised by Newton in his *Principia* against the wave theory of light, he continues (pp. 475–476): "J'ay crû devoir aller au devant de ces objections que pouvoit suggerer le Livre de Mr. Newton, sçachant la grande estime qu'on fait de cet ouvrage, & avec raison; puis qu'on ne sçauroit rien voir de plus sçavant en ces matieres, ni qui temoigne une plus grande penetration d'esprit."
88. The premises of the argument can be stated as follows: (1) "By the third law," gravitation is mutual. (I put the initial phrase in quotation marks because this is by no means a straightforward application of the third law: the latter states that the actions of bodies upon one another are equal and opposite; but it does not imply that if, for example, the moon is urged by a force *toward* the earth, that force is exerted on the moon *by* the earth—an implication that Newton explicitly disavows for his use of the term *attraction* (see *Principia,* I, Section XI]; and if, e.g., as Newton considered possible, the force of gravitation were in fact exerted by some sort of "ether" in contact with the gravitating body, the reaction required by the third law would be exerted *upon that ether,* not upon the body toward which the force is directed.) (2) Gravity—that is, weight—is a force exerted, not simply "upon a body" *as a whole,* but upon *each particle* of that body, in proportion to the mass of the particle (or: throughout the volume of the body, with a force-density proportional, at each point, to the mass-density of the body at that point). It follows from these premises that the gravitation is mutual not simply between earth and moon, or sun and planet, but between *each particle* of the one and *each particle* of the other. (3) The force of gravity toward any center of that force is characterized by a centrally directed *acceleration-field*—in Newton's terminology by a distribution throughout space of a "centripetal force" whose "accelerative measure" is independent of the body acted upon, depending only on the position of that body in relation to the center of force; more precisely, the accelerative measure is inversely proportional to the square of the distance of the body from the center. (This premise is based, so far as the law of the inverse square is concerned, upon the conclusions of the first phase of Newton's argument; and so far as the *universal* application to *all* bodies is concerned, on the known laws of terrestrial weight—and the further consideration that any particle in the universe can in principle become part of a terrestrial body (see *Principia,* III, Prop. VI, Cor. 2), so that if there were kinds of matter not equally subject to gravitational acceleration, one would expect to find terrestrial bodies that accelerate unequally in free fall.) It now follows directly that one must attribute not only "gravitational susceptibility," but also "gravitational power," to all bodies; and from the "accelerative" nature of the gravitational field, that the ("motive")

force of gravitation between two bodies is proportional to the mass of each. Since this force is also inversely proportional to the square of the distance between the two, and since no other factor influences its value, the law of universal gravitation in its complete form has been attained.

89. Howard Stein, "Newtonian Space-Time," *Texas Quarterly* (1967), 180; also in Robert Palter, ed., *The* Annus Mirabilis *of Sir Isaac Newton* (Cambridge, Mass.: MIT Press, 1970), p. 264.

90. Newton, *Opticks*, p. 397: "The *Vis inertiae* is a passive Principle by which Bodies persist in their Motion or Rest, receive Motion in proportion to the Force impressing it, and resist as much as they are resisted."

91. Alan E. Shapiro, ed., *The Optical Papers of Isaac Newton*, vol. I (Cambridge: Cambridge University Press, 1984), 86/87–88/89 (Latin and English, respectively; I have slightly revised the translation in my text).

92. Newton, *Opticks*, pp. 401–402.

Chapter 3

Foils for Newton: Comments on Howard Stein

Richard Arthur

It might have been supposed from the title of Howard Stein's paper that its central concern would be Locke's indebtedness to Huygens and Newton. Indeed, I pored over the first two installments he sent me to try to detect what line of influence he would trace: perhaps he would find the source of Locke's conception of *powers* of objects in Newton's reference in the preface of his *Principia* to his forces of inertia as *natural powers* of bodies; perhaps he would find the origins of Newton's laws of motion—which in his early work are the laws governing the translation of regions of space endowed with impenetrability—in Huygens' program of basing all of physics on the laws of impact; perhaps again Locke's fastening onto *impenetrability* (or *solidity*) as the defining characteristic of matter, which distinguishes it from space, would be traced to the corpuscular metaphysics shared by Huygens and the young Newton.

These questions of influence, however, despite their intrinsic interest, are clearly not the central concern of his paper. Indeed, if we approach it with the expectation that it is intended to throw light primarily on Locke and his sources, then it takes a rather strange line of attack. It begins with what appears to be a digression on the inadequacy of Cartesian physics, and how Huygens had first demonstrated the falsity of Descartes's laws of impact, then continues with an elegant reconstruction of how Huygens erects on the ruins of the Cartesian theory many of the principal results of "Newtonian" physics by an ingenious application of Galileo's principles. Then, nearly halfway through the paper, Stein turns rather abruptly to a consideration of Newton's theory of light and the methodology that accompanies it, and proceeds to a comparison and contrast of Newton's views on primary and secondary qualities and the nature of scientific methodology with those of Locke. Here finally we have Locke, though not as following from Newton but rather as contrasted with Newton, whose views on methodology are then in turn contrasted with those of Aristotle and Huygens.

But the paper's ending gives us the key to its interpretation. It is not about Locke or his sources; its centerpiece is Newton, and Locke and Huygens both are merely *foils*. Thus the primary motivation for the discussion of Descartes's and Huygens' physics is not, I believe, for their intrinsic interest—although this is very great, and Stein is as usual succinct and enlightening—but for the contrasting light this throws on Newton's method, and the place of metaphysics within his thought. In other words, this is another in a series of erudite papers by Stein, whose intention is the rehabilitation of Newton as a philosopher—metaphysician, methodologist, and epistemologist—of the first rank.

I do not mean to imply by these remarks that Locke and Huygens are unimportant to Stein: far from it. In fact the massive underestimation of Huygens' role in the formulation of what we call Newtonian physics forms another major theme of this paper—the *subplot*, as it were. Again I see this as part of a continuing project of Stein's, this time the rehabilitation of *Huygens*, the goal being to establish him as one of the great triumvirate of natural philosophers of the second half of the seventeenth century, along with Newton and Leibniz.

Indeed, it might almost be claimed that Huygens' role in founding modern physics is even more essential than that of his two famous successors. For it is Huygens who does the vital work of extracting the wheat from the chaff in the work of Galileo and Descartes, and forging them together into one highly coherent whole. It is, after all, Huygens who first formulates correctly what we know as the principle of Galilean relativity; who learns the lesson that Galileo had to teach about the "nonoperative" character of motion, and forges this together with Descartes's and Gassendi's formulations of the law of inertia into the fully modern concept; and it is Huygens who has a full theoretical framework for understanding the law of conservation of momentum, established experimentally by himself as well as by Wren, Wallis and Marriotte. But in this paper Stein takes us further and shows how, in a "very beautiful argument," Huygens derives the law of conservation of momentum from the principles of inertia and relativity, together with the third (Torricellian) principle that "by a motion of bodies that results from their gravity, their common center of gravity cannot be raised." This certainly is a beautiful argument; and it is quite exhilarating to follow Stein in seeing how these same graceful and economic reasonings of Huygens lead him to his expression for the centrifugal force, his distillation of the concept of mass and its correct distinction from weight, and on to the brink of Newton's second law.

One thing that made a particularly strong impression on me was Stein's observation that Newton—very modestly—makes no claim for the originality of his three famous laws; and this despite the fairly "radical shift" in interpretation away from Cartesian impact physics on which Newton had already embarked by the time he wrote them.[1] His implied acknowledgment of Huygens as one of the "Mathematicians" who had "receiv'd" these principles thus becomes particularly intriguing in the light of Stein's demonstration of how close Huygens came to having the essential contents of Newtonian physics many years in advance of Newton.

Nevertheless it seems to me that the primary function of this semidigression on Huygens' physics in Stein's paper is to emphasize that metaphysical foundations play a different role in Huygens' methodology than in Newton's, and moreover to show the superiority of Newton's methodology. As Stein argues, there is a considerable correspondence between Huygens' strongly mechanistic and corpuscularian philosophy—"the true Philosophy, in which one conceives the causes of all natural effects in terms of mechanical notions"—and Newton's early metaphysics of space and substance. For in Newton's "De Gravitatione" (generally dated from the late 1660s) a body or substance is (conjecturally) defined as a region of space that God has endowed with three properties: those of *impenetrability, movability,* and the *capacity to interact with minds.* The first two of these properties, according to Stein, show a commitment to the Huygensian view that all action among bodies is action by contact, these bodies being "absolutely hard" by virtue of their definition as impenetrable regions of constant shape and size. (Here, too, we have not only an obvious precursor of Locke's definition of body but also, Stein argues, a far superior and less obscure metaphysics of substance.)

But the contrast between Newton on the one side and Huygens and Locke on the other is in the fluid and nondogmatic manner in which Newton treats these metaphysical foundations. Unlike them, he does not subscribe to the conception of science as founded on *fixed* first principles, a conception inherited by Locke and Huygens from Aristotle and Descartes. Huygens, Stein tells us, is so deeply committed to the corpuscularian metaphysics, to the reduction of all causes to mechanical ones, that he sees no other way for physics to progress. And Locke is equally dogmatic about our inability to conceive any other way for bodies to operate on one another—or at least on us—than by impulse, whatever his admissions to Newton's influence.

Newton, by contrast, is quite willing later to abandon his commitment to the purely mechanical principles of impact physics. For he does not believe that first principles can be established prior to the investigation of the phenomena. Although he agrees with the demonstrative or deductive ideal of scientific knowledge—how could he not, as one of the supreme geometers of all time?—the principles from which it is to be deduced are, as Stein observes, "tentatively first" or "relatively first" principles. But this does not mean that they are hypothetical *causes;* hypotheses are attempts to give models or explanations of the causal processes underlying the principles, but they are not these principles themselves. Moreover, Newton's principles are to be found empirically; they are to be "deduced from the phenomena."

It is noteworthy here that in three important respects—the denial that first principles or causes can be known with certainty, the dependence of knowledge on "tentatively first" principles, and the insistence that these must be established empirically and not by rational intuition—Newton follows Descartes's great rival, Pierre Gassendi. Of course, what Gassendi achieved in scientific practice falls far short of what he preached on method. Nevertheless, a full contrast between Gassendi and Newton would do much to illuminate the originality of Newton's method, which lies particularly in his strong emphasis on mathematics. For one major respect in which Newton differs from Gassendi is in his claim that the principles he deduces from the phenomena are *mathematical* principles, not explanatory hypotheses.

This is even the case, as Stein observes, in Newton's optical investigations, where he deployed this original conception of scientific method to the "bafflement" of Hooke and Huygens. Principles such as that *"Light* consists of Rays differently refrangible" were claimed by him to be nonhypothetical. Neither the splitting up of light into component rays by means of a prism nor the measurement of the refractive index of each of these rays depended on any prior hypothesis as to the nature of light. Yet the property of refrangibility is wholly mathematical, each ray having a unique and invariable index; and the fact that each ray has such a property is deduced from the phenomena. Furthermore, even though he insists on the nonhypothetical status of these principles, Newton is quite clear, both in his reply to Hooke's objections and in his official Rules of Reasoning, that they are *revisable.* This is seen by Stein as evidence that Newton's scientific method is inherently *dialectical: "forward* in the mastery of phenomena, and *"backward"* in the discovery of new principles as a *framework for further investigation* (rather . . . than [as] a "necessary foundation" for physics),"[2] as it is in Huygens and Locke.

I have some reservations here, however. In support of his reading of Newton's method as dialectical, Stein invites us to see Newton's abandonment of impact physics as an example of his flexibility on first principles. But, in the first place, it is not so clear to me that Newton was as committed to the Huygensian metaphysics in his youth as Stein would have us believe; nor, second, is it obvious that the undoubted development of Newton's metaphysics involves any overturning of principles. Indeed, as to the first of these criticisms, I would say that, inasmuch as he never did believe that ultimate natures or causes could be finally discovered, Newton always remained something of a Gassendist, and never had more than a tentative commitment to the various causal models he considered. Thus, granted that his early definition of body and laws of motion are compatible with a mechanistic physics of action by contact, this is no less true of the mathematics of the *Principia,* although by then Newton had despaired of finding a mechanistic cause of gravity. Moreover, even in Newton's early tract "De Gravitatione," in the passage alluded to by Stein in which he gives his fictional account of God's creation of matter, the regions of space are granted not only the "mechanical" properties of impenetrability and movability but also the highly antimechanical one of being able to act on and be acted on by minds. This last property, of course, was never far from the front of Newton's own mind, and was always the one he appealed to in defense of the intelligibility of action-at-a-distance.[3]

Second, although Newton's metaphysics undoubtedly undergoes development, this does not seem to be as the result of the overturning of mathematical principles. It is true that, by the time of the *Optics,* Newton had come to reject explicitly the Huygensian program of reducing all interaction to the impact of absolutely hard bodies, and to propose the introduction of new principles, what he calls the *active principles* of gravity, fermentation, and cohesion. But these principles do not replace but rather complement the old *passive principles,* the passive laws of motion which result from the force of inertia. Again, these active principles are considered to have the status of "general Laws of Nature," and to be derived from the phenomena, "though their causes be not yet discovered."[4] This sets them apart from any hypotheses designed to explain them, any causal models, which, for Newton, would stand or fall independently of them; whereas emendation of the principles themselves would require a more complicated process of criticism, involving a revision of the experimental evidence itself.

But if this is so, then is Newton's method really any more "dialectical" than positivistic accounts of scientific method of a more recent vintage, for instance Nagel's in his *Structure of Science*? In some respects I think it is not. For although Newton did not claim absolute certainty for the principles he derived in his ray optics, he was no more prepared to acknowledge that such principles could be supplanted on the basis of a rival theoretical interpretation of the phenomena, such as Huygens' wave theory, than Nagel would acknowledge the susceptibility to theoretical criticism of an "empirical law" such as Balmer's law.[5] As he stated in one of his draftings of Rule 4, "If a certain proposition drawn by induction is not yet sufficiently precise, it must be corrected not by hypotheses, but by the phenomena of nature more fully and more accurately observed."[6] Again, notwithstanding his concessions to their revisability, the fact remains that Newton framed his mechanical principles with such consummate skill and caution that neither he nor anyone else managed to emend them significantly for two hundred years. We would do well to remember, I think, that nothing did more to engender positivism than this combination of the seeming impregnability of Newton's physical principles with his claim to have strictly derived them from the phenomena.[7]

Yet in respect of his handling of *metaphysical foundations*, Newton is, as Stein points out, a good deal more flexible and less dogmatic in his methodology than either seventeenth-century corpuscularian philosophy or recent positivism, for his metaphysical views are in constant dynamic interplay with the mathematical principles they are designed to explain, as well as with the phenomena derived from them. Even though causal hypotheses may not be used to found principles, they constitute the heuristic within which the principles are discovered, and are then evenhandedly modified or rejected by Newton solely on the basis of their compatibility with these principles and the phenomena. A case in point is his theory of substance, whose development is deftly traced by Stein from its implicit beginnings in "De Gravitatione" through to the mature exposition in the *Optics*. Although it is always tentatively proposed, Newton's conception of a substance as a region of space endowed with the powers of impenetrability, movability, and the power to interact with minds is modified as the principles underlying these properties—the "general Laws of Nature"—are progressively unfolded within his research program as a whole.

This dialectic is also discernible in those aspects of Newton's metaphysics that he does not present as hypothetical or tentative, for instance his concepts of space and time. Stein himself has already shown elsewhere how Newton's concept of absolute space, far from being ineffable, is developed in the process of some penetrating criticisms of Cartesian metaphysics and its inability adequately to represent motion. The same is true of his concept of absolute time, as I have tried to show in some recent work of my own.[8] Of its supposedly otiose properties—its flow, the equableness thereof, its distinction from relative time—none is without some counterpart in the specifics of Newton's mathematical physics. The flow of time, for instance, is based in Newton's fluxional mathematics, whereas the equability of its flow is related to Newton's proof of the existence of absolute motion, and is thus consequent on the success of the program of the *Principia*. Thus in Newton's sophisticated empiricism, unobservable and metaphysical elements are comfortably integrated into the whole empirical program.

Locke, of course, when used as a foil for this sophisticated empiricism of Newton's, comes off much the worse for the contrast. He compromises his official commitment to action-by-contact metaphysics, without indicating how his "official position" (Stein's term) could survive the admission of action-at-a-distance. He conflates Aristotle's two kinds of first principles, best known in nature, and best known to us, thus setting under way a long empiricist tradition of confusing epistemology with ontology, which is evident also in his empiricist derivation of the primary qualities. And in setting bounds to human knowledge in the capacity of human perception, Locke ends up denying the possibility of a scientific natural philosophy (albeit with unassimilated qualifications).

These conclusions are particularly ironic, of course, given the usual treatment of Newton as a philosopher. For it is generally Newton who is taken to task for his dogmatism in metaphysics,[9] and for "personally taking leave of his empiricism."[10] Newton bashing has become a popular tendency among philosophers in this century. In part, no doubt, this is an instance of that same psychological phenomenon that produces all the cranks who would refute Einstein's relativity. But more excusably it is an inevitable outcome of the deification of Newton, or at least his papification as head of the Church of Positivism. This at any rate explains Burtt's hostility, and perhaps Reichenbach's too. Both see a gap between the "official philosophy"—the "deduction from the phenomena" and the "hypotheses *non fingo*"—and Newton's actual practice, which involves the entertainment of hypotheses, as well as starkly metaphysical elements such as his concepts of absolute space and time.

But, as Professor Stein's paper has done so much to demonstrate, such judgments reflect more a failure on the part of his accusers to analyze Newton's arguments with sufficient care and diligence, as well as a failure to analyze his actual scientific practice. This failure too, I would suggest, has more than a merely accidental connection with the topic of this essay. For the usual professional initiation into philosophy as a discipline distinct from science is through a study of

Descartes, Locke, Berkeley, and Hume, which tradition has also done most to promote the end of philosophy as a constructive, interactive contribution to science—the kind of philosophy we find in Huygens, Newton, and Leibniz. The divorce of epistemology from the main fount of new knowledge—science—begins with Locke's pose of subservience to "the incomparable Mr. Newton" and issues in the contempt for Newton's philosophical competence expressed by scholars such as Reichenbach and Burtt. The lasting value of Howard Stein's paper, I suggest, will be in restoring our perspective on the relative merits of Huygens, Newton, and Locke by showing us what we still have to learn from Newton the philosopher.

Notes

1. For a detailed analysis of the development of Newton's concept of force out of impact physics, see Alan Gabbey, "Force and Inertia in Seventeenth-Century Dynamics," *Studies in the History and Philosophy of Science* 2, no. 1 (May 1971), 1–67.
2. This wording—slightly more elaborate than that given in his paper—was used by Professor Stein in a letter to me, dated October 8, 1987.
3. Newton's reasoning would run as follows: Since on this account a body is "nothing more than an *effect* of the Divine mind elicited in a definite quantity of space," and God is able to act through his will directly on our perceptions, there is no reason why he should not adjoin the same power to the effects of his will—i.e., to bodies. Similarly, since God can act anywhere and everywhere, there is no reason why the effects of his will should not also.
4. *Optics*, p. 401; quoted from Stein's "On Metaphysics and Method in Newton," a sister paper to the one presented by Professor Stein, also in manuscript form, which he kindly sent me while I was working on my comments; p. 42.
5. For a criticism of Nagel's philosophy of science, see Paul Feyerabend's review in *BJPS* 16 (1964), 237 ff.; reprinted in volume 2 of his Philosophical Papers, *Problems of Empiricism*, pp. 52–64; see also my "The Empiricist Account of Scientific Knowledge: A Polemical Evaluation," *Poznan Studies* 3, no. 2 (1977).
6. Quoted from Feyerabend's "Classical Empiricism," pp. 150–170, in *The Methodological Heritage of Newton*, ed. R. E. Butts and J. W. Davis (Blackwell, 1970); reprinted in volume 2 of his Philosophical Papers, *Problems of Empiricism*, pp, 34–51; p, 42, n. 7.
7. For criticism of Newton along these lines, see Feyerabend's classic article "Classical Empiricism," op. cit.
8. "The Metaphysical Basis of Newton's Method of Fluxions," presented at the History of Science Society Annual Meeting, Raleigh, North Carolina, November 1, 1987.
9. Cf. Reichenbach's assessment of Newton as a "physicist who, whenever he left the domain of his narrower specialty, became a mystic and a dogmatist," quoted in Stein's "Newtonian Space-Time," *The Texas Quarterly* (Autumn 1967), 174–200, 188–189.
10. Edwin A. Burtt, *The Metaphysical Foundations of Modern Science* (Doubleday Anchor, 1954), p. 244.

Chapter 4
Real Quantities and Their Sensible Measures
Lawrence Sklar

Wherefore relative quantities are not the quantities themselves, whose names they bear, but those sensible measures of them (either accurate or inaccurate), which are commonly used instead of the measured quantities themselves. And if the meaning of the words is to be determined by their use, then by the names time, space, place, and motion their sensible measures are properly to be understood; and the expression will be unusual, and purely mathematical, if the measured quantities themselves are meant. On this account, those violate the purity of language, which ought to be kept precise, who interpret the words for the measured quantities. Nor do those less defile the purity of mathematical and philosophical truths, who confound real quantities with their relations and sensible measures."[1]

i

Thus Newton sums up the moral to be drawn from his brief but deservedly famous Scholium to the Definitions of the *Principia*. The Scholium, with its defense of "absolute, true and mathematical time," which "of itself and from its own nature flows equably," and "absolute space," which "in its own nature, without relation to anything external, remains always similar and immovable," is, of course, the key text contra relationism, a vital component of the Newtonian argument which converted the debate between relationists and substantivalists about space and time from one grounded purely on familiar philosophical considerations to one in which experimental and empirical elements were alleged to play a crucial role.

Despite the vast amount of attention brought to bear on the arguments of the Scholium and on the derivative arguments that have followed from it attempting to refine its thesis, bring them into harmony with later physical science, and then buttress their conclusions or reveal their fallacious nature, I think there may still be some things to be learned, both historically and philosophically, by once more going over the familiar, but sometimes perplexing, original text.

Let me begin by outlining briefly the structure of the Scholium.[2]

First it is asserted that time, place, and motion are commonly conceived of as relative to sensible objects. Next, absolute time "flowing equably" is contrasted with the more or less accurate time revealed by sensible clocks. Absolute space is then defined as the immovable frame relative to which absolute motion is determined, and it is contrasted with the mere relative place and motion of material reference frames. Place is defined as that part of space taken up by a body, and absolute motion as "translation of a body from one absolute place to another."

This statement of motifs is followed by a "development" section. In astronomy the equation of time is used to correct our ordinary time measure for doing astronomical dynamics, a typical example of our understanding the distinction of a real time interval from its only more or less accurate "sensible measure." As the "order of the parts of time" is immutable, so is the "order of parts of space." But parts of space itself "cannot be seen, or distinguished from one another by our sense," so instead of absolute space we normally rely on mere relative place and motion "without any inconvenience in common affairs." But such equivocation is dangerous in "philosophical disquisitions," where the distinction of absolute and merely relative place is essential, "for it may be that there is no body really at rest, to which the places and motions of others may be referred."

A body included in another is at rest with respect to the whole of which it is a part, but it may be in absolute motion. So again absolute motion cannot merely be relative motion. Only change of absolute place constitutes absolute motion. And absolute places are, of course, immovable for all eternity.

If genuine place, rest, and motion are not discernible by our sense, how can they be known? The answer is, of course, that it is by causes and effects that you shall know them. Here the famous spinning bucket experiment is introduced. Relative circular motion of water and bucket does not generate the forces responsible for the surface of the water deviating from flatness. And while there are a variety of relative circular motions of water to material objects, it is only the one true circular motion that is correlated with the generation of the inertial forces. As is true on earth so is it in the heavens, and celestial objects, insofar as they partake of genuine circular motions, suffer the inertial effects thereof.

From this the conclusion that initiated this paper is drawn: Relative quantities are not the quantities themselves but only sensible measures of them. And to confound the two by equivocation of terms is a mistake in both directions.

Following this comes a short discussion of how we might actually determine the true motions of bodies. One guide is that apparent motions are the differences of true motions, which is, presumably, important since it means that if one true motion is established, from then on the others can be determined by use of relative motion alone. And it is forces that can guide us to establish the true motion of some reference object. Here the two globes on the ends of a rope are introduced, with the measured tension in the rope indicating their true direction and magnitude of rotation about their common center. Their absolute circular motion is then determinable "even in an immense vacuum, where there is nothing external or sensible with which the globes could be compared." In our universe we could determine even the true motion of the fixed stars by the use of some such dynamic indicator of true motion to serve as reference frame fixer and the relative motion of stars to such an object whose true motion has been dynamically determined.

ii

One "quirk" of the text which immediately strikes the contemporary reader is its emphasis on real circular motion as the paradigm of absolute motion to the total neglect of the equally compelling facts about the inertial effects of linear accelerations as components of an argument for absolute motion and absolute place. One result of this is that, at least in this narrow context, the puzzle of the empirical indeterminability within the Newtonian frame of uniform motion with respect to space itself is hidden from the reader's view, although, of course, Newton is intimately familiar with the consequences of Galilean relativity, as Corollary V of the Axioms plainly shows.[3] Pursuing this line leads to the well-explored path of seeking a space-time that does justice to the Newtonian empirical facts about the dynamic consequences of absolute acceleration but has no place for absolute uniform motion, a topic I will forgo here.

Let us look at some of the contrasts Newton draws between the "quantities themselves" and their "sensible measures." Quantities themselves are what are referred to in scientific discourse. To misunderstand the terms of mathematical and philosophical discourse (which I think we can in current terms think of as scientific discourse) as referring to the sensible measures is to "defile the purity" of the scientific assertions. The quantities themselves are not available to the senses but are known to us only by their causes and effects—that is, by causal inference from the behavior of the sensible measures available to our sensory inspection. The sensorily available phenomena are related to the quantities themselves as their *measures*. As such they can be accurate or inaccurate measures of the quantities themselves. The sensory measures are relative quantities. They are, explicitly or implicitly, relational in nature. But the quantities themselves by contrast have some kind of absolute nature, understood as monadic or nonrelational.

But if we try to apply all of these contrasts to the quantities and contrasted measures cited by Newton in the Scholium, we are forced to say some things in expounding his view that at least *sound* peculiar. On the one hand, does it seem right to say that the genuine temporal interval between events is known to us by causal inference from its causes and effects? On the other, isn't it at least a little peculiar to say that the relative motion of objects, one to another, is somehow or other a more or less accurate measure of genuine or absolute motion, assuming with Newton that there is such a thing? Are the peculiarities of expression here just that, or do they, perhaps, when reflected on, lead us to some deeper underlying issues?

We can trace the origin of this at least verbal peculiarity back to the first statement of the theses in the Scholium. Absolute, true, and mathematical time is contrasted with the more or less accurate measures of it provided by periodic material processes. But absolute space is not contrasted with the more or less accurate measures of it provided by material measures of the spatial metric—tape measures, yardsticks and the like—but rather with material reference frames relative to which the magnitude of *motion* is to be determined.

Now, the reason for this contrast in the way time and space are dealt with is, I think, partly clear, but partly a matter for our speculation. The Scholium to the Definitions is immediately followed in the *Principia* by the Axioms, the statement of the laws of motion. The central law, the second, equates impressed force with change of motion. Were time to be plugged into the equation using more or less inaccurate clock time not corrected to true elapsed time, or were change of motion to be construed as change of motion determined in some reference frame not in itself inertial, the rate of change of motion to which impressed force is proportional would be incorrectly calculated. Add to this two facts: first, that it is astronomical data which provides the overwhelming bulk of the data confirming the conjunction of the second law with the inverse square law of gravity and the consequences of these which are the primary subject of the *Principia*, and second, that the correction of solar time, which is variable owing to the variable speed of the earth in its orbit and the inclination of the equator to the ecliptic, to more uniform elapsed time by means of the equation of time plays an important role in astronomical calculations, a practical role not matched by any similar systematic need to correct for distance measurements in this way. So one can see why it is important to emphasize the need for theoretical, mathematical, and absolute time and theoretical, absolute, and mathematical motion as the time and motion meant when these terms are invoked in the Axioms that follow. Failure to recognize the need for taking the terms used in this way and confusion of absolute time or absolute motion with their "sensible measures" would result in the purity of the mathematical and philosophical truths being "defiled." We would be plugging the wrong quantities into the equation of motion.

Be that as it may, it is curious that Newton did not pair an "absolute metric" of space with its more or less (in the form of measuring rods) inaccurate measures as the parallel to the absolute metric of time and its more or less inaccurate measure (in the form of periodic processes serving as clocks), but instead moved directly from time itself to space itself rather as the reference relative to which motion constitutes absolute motion, and to the quantity of "motion itself" as opposed to mere relative motion with respect to material objects.

It is in dealing with absolute time as contrasted with time determined by clock readings that we are most comfortable with Newton's contrast between quantities themselves and their mere sensible measures. Between a pair of events there is the actual temporal interval. We can estimate this interval by means of some periodic process, but such an estimation will always be subject to the variety of inaccuracies from which a measurement process can suffer. It will be part of the aim of our completed physical theory to explain why such errors of measurement occur, allowing us to construct ever better clocks that more accurately reveal to us the real time intervals between events.

Newton is certainly on to something of vital importance here. Whatever our metaphysical views about space and time—relationalist or substantivalist, representationalist, fictionalist, or empiricist— whatever we take space and time and reference to them to be, we must all agree that there are natural measures of space and time. When we represent the time interval between events properly, the laws of nature take on elegant and simple forms. When we fail to get this measure right, whether it be the measure of absolute time in Newtonian or neo-Newtonian space-times or the affine parameter measuring proper time intervals along timelike paths in relativistic space-times, we cannot find the simple laws involving the t parameter which so compactly and elegantly represent the world when we have the time measured right. And it is a fact of nature—certainly one of its most profound and important facts—that a wide variety of clocks (atomic, nuclear, or those using light rays bounced back and forth between mirrors carried by inertial particles) will not only agree with one another in their metric of time—subject, of course, to the greater or lesser deviation that is the fate of all merely sensible measures of time itself—but will measure time in a way that at least approximates the natural measure. And the same goes for measures of spatial distance and its more or less accurate measurement by material measuring rods.

Indeed, there are other aspects to these "natural measures" that go beyond their suitability for elegantly formulating general laws of nature. In statistical mechanics, both equilibrium and nonequilibrium, we are required to posit a distribution of the initial conditions of systems over the possible initial dynamical states allowed the systems by the constraints (being confined to a box, having constant energy, and so on) placed upon them. The distribution chosen is always some variant or another of the distribution obtained by taking "all regions of possible initial conditions of the same size" as equally likely to contain the initial state. But what is it for a phase-space region of a certain set of initial conditions to be the "same size" as another such region? To make this a well-defined notion requires that a measure be placed over the phase-space. This is just the instantiation in statistical mechanics of the well-known general problem of applying a principle of indifference or of symmetry to obtain a priori probabilities. Without a measure imposed over the space of possibilities, a principle of indifference is without genuine content.

The measure that "works" in statistical mechanics is the natural measure generated from our natural measures of space and time. Actually it isn't as simple as that, for while the spatial component is our ordinary spatial measure, the other component must be the measure of the appropriate "quantity of motion," which, it turns out, is momentum (rather than energy, for example). But the measure chosen is at least dependent on our first choosing the natural measures of time and space that work so well for framing the simple laws of dynamics. To a degree, at least in equilibrium statistical mechanics, we can, by means of reliance on ergodic theorems, offer a rationale for the choice of the natural measure of phase-space which ties that measure into the role played by the natural measures of time and space in formulating the simple dynamical laws. But in the nonequilibrium case it is not clear that this can be done in any persuasive way. So here we seem to have a ground, over and above the argument from the simplicity of laws framed in terms of time and space as measured in the usual way, for saying that the "real" measure of time and space exists as an ideal over and above that which we can measure with clocks and rods, and for once again feeling the pull of

Newton's distinction between the quantity itself and its merely sensible measure.

But what is it for our clocks and rods to be sensible measures of time and space themselves? If we were dealing with spatial measure, which, as we have seen, Newton ignores, the picture of laying off our meter sticks against some standard meter to determine its accuracy as a measure comes to mind, and of then using the meter stick to determine more or less accurately the distance between points correctly measured only by that standard meter. But here it is supposed to be the true distance between the points that is the standard itself, not some comparison of one sensible measure with another.

Understanding the Newtonian language here is replete with problems. In what sense is it the case that the distance between points, the true distance, is itself not sensibly available to us? In what sense is it the case that this distance is known only by its causal effects? Does the true distance between the points somehow cause the meter stick readings that we take as a measure of that distance to have the values that they have?

When we move from space to time, things seem even more peculiar. In what sense is the temporal interval between events insensible to us? How are we to understand the measurement relationship between the intervals ticked off between events on our more or less accurate clocks and the true time interval being measured? Here the metaphor that I used earlier of laying one meter stick against another seems even less appropriate. It is bad enough to think of space itself as a kind of thing, like a meter stick or table against which comparisons can be made. Thinking of time as a kind of quasi substance in this way seems even less intuitive to us. Conceptual confusion here is made even more likely, of course, by Newton's resort to the familiar but horribly confused metaphor of time as something that "flows equably," with its misleading suggestion of time both as something substantial and as something to which a "rate of motion" can be attributed. Again we also want to ask about the general suggestion that the philosophical and mathematical elements are known to us only by causal inference from the behavior of the sensible. Does the actual time interval between the events explain our clock ticking measurement results as cause explains effect? Is the *cause* of the clock (atom, nucleus, bouncing light ray, or whatever) ticking off such and such a value between the events the existence of such and such a time interval between the events?

Surely what has happened here is this: The two different problems of the true metric of space and time versus their more or less accurate measures by material measuring instruments, and of the absolute motion of objects versus mere relative motion of an object to some arbitrary material reference frames, are being treated together. At this point the language appropriate to the latter problem—that is, the unavailability of the true quantity to direct sensory apprehension— and the evidence for it consisting in its causal effects in the realm of the material have been applied to the former, and thus a great deal of strain has been placed on our intuitions about language, about ontology, and about what counts as a legitimate causal relationship.

The peculiarities of language that arise when we treat the two problems of absolute time and absolute motion together appear even more striking from the other side of the contrasting pair. It does seem reasonable to talk of absolute motion, as opposed to relative motion to a material frame, as insensible, as not available to immediate sensory detection. And it seems reasonable here, given Newton's views on the matter, to speak of absolute motion as known to us by its causal effects—that is, by the sensorily available responses to inertial forces generated by acceleration with respect to place itself, ignoring for the present, of course, the crucial problem that only absolute acceleration and not absolute velocity (which one would think of as absolute motion if anything is) is so revealed by its causal consequences.

But in what way is relative motion a sensible measure of absolute motion? In what sense is it plausible to assert that relative motions provide some more or less accurate measurement determination of the underlying quantity absolute motion? Now, it is true, as Newton emphasizes, that relative motions are the differences between true motions. So if we once know the true motion of some material reference frame—say, the famous weights on the end of a string whose true circular motion is determined by the tension in the rope—we can then use the more or less accurately determinable measure of the relative motion of other objects to the chosen reference standard to determine the true acceleration of these other objects. But that is quite different from the way in which clocks and rods are taken to provide more or less accurate measurements in themselves of the absolute metrics of time and space. It is the evidence of the inertial effects, if anything, that is the measure of absolute motion, not relative motion. So while Newton can consistently maintain that the structure of space and time that captures the notion of absolute rest, and hence of real as opposed to relative motion, is, like the structure of space and time that captures the notion of real as opposed to merely apparent metric interval, "insensible" and known to us only indirectly through our sensible awareness of the behavior of material objects, his assimilation of relative motion to clock readings, the one being the "sensible measure" of absolute motion as the other is of temporal interval, is at least peculiar as a mode of expressing what he wants to say.

iii

At this point it might seem that the peculiarities of language I have noted are just that: peculiarities of expression with no deeper import. I do not think that is quite right, although my reason for saying so will have to follow an indirect route.

From our contemporary pespective we have all the systematic means at our disposal for saying in a comprehensive and rather unified way what Newton wanted to say. Indeed, we can generalize from what he had to say to a general way of presenting space-time theories of a "substantivalist" sort which carry on the Newtonian tradition into the realm of relativistic space-times or the curved space-times of contemporary gravitational theories. We posit a general space-time structure, be it Newton's, flat neo-Newtonian space-time, the curved neo-Newtonian space-time of reconstructed Newtonian gravitational theory, the flat Minkowski space-time of special relativity, the curved pseudo-Riemannian space-time of general relativity, or any other variation or generalization on these we come to find useful in our theorizing. The space-times themselves and their structural features remain Newtonian insensibles, not directly open to our observational determination. Our knowledge of what space-time is like is always indirect, inferred by some process of theoretical postulation from the sensible behavior of the material inhabitants of space-time: rods, clocks, free particles, or light rays. Not, of course, that we are claiming these or their behavior to be in some sense perceptually immediate or theoretically uninterpreted, but only that they stand at least one step of the ladder between the insensible space-time features and what is empirically available to us.

The space-times can have many different structures: metric intervals along one-dimensional paths in them (spatial, temporal, or spatiotemporal), conformal structures of angles, projective structures and affine structures determining local parallelism of tangent vectors and the subclass of one-dimensional paths which are geodesics, topological structures and manifold structures underlying these, and so on. These may be independent of one another in various ways, or, like the metric and affine structure in pseudo-Riemannian space-times, they may be interdefinable in varying degrees and directions. Fundamental to our theory will be the laws connecting space-time structure to the behavior of material objects. In general relativity, for example, we will need things like ideal atomic clocks to measure proper time intervals along the timelike paths of their life history, light rays *in vacuo* to follow null geodesics, free particles (at least spin-free monopoles) to follow timelike geodesics, and so on. It is, of course, these "bridging postulates" that connect what we can see and measure among the sensible material with what we can posit as the structure of the insensible space-time itself. Put this way, the connection of the material with the spatiotemporal can be uniformly described, and we need not worry about the fact that it sounds peculiar to say the intervals of time "cause" the clocks to read as they do, just as it sounds peculiar to talk of relative motion as some "measure" of absolute motion.

I am not here arguing that this is the right way to think about space-time, the material world, and their relation to each other, or even that this way of talking about things is not fraught with problems, but only that it seems to be the natural way of presenting a Newtonian line brought up to date, generalized, and purged of linguistic peculiarities.

It is one familiar difficulty with this picture to which I now want to turn. The difficulty is already latent in Newton's original account of the Scholium. The origin of inertial forces is in motion (or rather acceleration) with respect to place itself, and the magnitude of such forces is proportional to the magnitude of this "real" acceleration. This explains why relative acceleration results in a difference of felt inertial force, since relative motions are differences in real motions. But what shred of evidence does Newton have for this theory? There is plenty of evidence suggesting that relative inertial effect is proportional to relative acceleration. And there is the fact that some systems feel (more or less) no inertial effect at all. But what evidence is there to the claim that the systems feeling no inertial effect are unaccelerated with respect to place itself? None, other than the circular grounds that on his theory such uniformity of motion with respect to place is what is to be correlated with no inertial effect being felt.

A world that had absolute place and in which zero inertial effects were felt by systems having some fixed nonzero acceleration with respect to place itself would, as far as sensible phenomena were concerned, look just like Newton's world. Indeed, a world without absolute place at all, but in which some systems simply felt no inertial forces while others felt inertial forces proportional to their acceleration with respect to these systems (or the reference frame of them even if there were no such actual systems in the world) would look the same as Newton's. It is in this latter direction that, in fact, relationists since Leibniz have, more or less opaquely, sought the correct relationist response to Newton's substantivalism. Note that the response here is not the move to more sophisticated space-times, such as the neo-Newtonian or relativistic ones, sometimes suggested as the resolution to this sort of problem. For they have the same sort of problem. In these space-times it is the association of inertial (or free optical) motion with the geodesics of the "space-time itself," which is as great a leap of faith (or convention) as Newton's association of inertiality with uniform motion with respect to absolute place.

Not, of course, that the substantivalist does not have a response to all of this, the best being some defense or other of the claim that the standard view (Newton's or his substantivalist successors') of the relation of space-time structure to observable effect is the "theoretically best" choice, where the standard of preference is based on some form or other of theoretical simplicity or parsimony.

The fundamental problem faced by Newtonian-like accounts of inertial forces in terms of absolute motion is faced also, on reflection, by Newtonian-like accounts of the metric behavior of material things as well. We can tell if our solar clock is running erratically relative to our pendulum clock, and if the latter varies in its time-keeping when scored by masers or by nuclear decay. And we can discover, staying in the realm of the sensibles, that there is a "corrected" time scheme, closer to that of the atoms than that of the sundial, relative to which time measurement gives rise to simpler laws of motion. But how, pray, can we tell that the ideal, "corrected" scale measures the true lapse of time itself, "flowing equably without relation to anything external"? Once again the substantivalist will have a reply for us, invoking some kind of legitimized inference to the insensible through some variant of simplicity or something like it. But our doubts remain.

iv

Now one way, although by no means the only one, in which the debate between substantivalists and antisubstantivalists has been formulated is to look at theories of space and time as they appear in their Ramsey sentence guise. The terms referring to sensible material objects and their interrelations are taken as primitive and understood as fully interpreted antecedent to the theory. The terms referring to space-time and its features are replaced by predicate variables quantified over by second-order existential quantifiers.

How are we to understand the existential commitments to space-time entities and their features asserted by the theory? According to the substantivalist, this commitment is to be understood concretely, as a commitment to the realm of entities and features of space-time itself, a realm of being in the world additional to the being of material objects and their features. According to the antisubstantivalist, in this version of that general stance, the quantifiers can be understood, rather, as referring to abstracta. These are the appropriate sets and numbers that one needs to fill in the existential commitment without taking too seriously some mysterious realm of insensible concreta. According to this view, we can have the predictive value of our theory, keep its compact formalization (instead of moving, say, to the set of all its "observational consequences" as a surrogate for the original theory with all the "messiness" of formalization that would entail), and yet realize how innocent of genuine ontological commitment all our talk of space-time and its features is.

For very good reason those who frame the debate in this way usually think of Newton as the prime exponent of taking the apparent ontological commitment of the theory seriously, and refusing to explain it away as mere introduction of abstracta into the picture, no more to be taken seriously as telling us that beings exist than would be a statement to the effect that "there is a coordinatization of places on the earth such that . . ." would introduce any more being into our picture than the earthly places themselves. After all, if Newton does not believe in the reality of space and time themselves, over and above the material inhabitants of them, who does?

But not everyone has read Newton in the orthodox "realist" way. Some years ago Stephen Toulmin published a highly interesting piece on Newton in which he argued, among other things, that reading Newton as espousing the independent existence of space and time in the Scholium might be a mistake. Toulmin did not put the contrast as one between realist and representationalist accounts of theoretical ontological commitment, but many of the things he says would, I think, fit nicely into a framework in which one read him as asserting that much that Newton says can be understood in a representationalist vein.[4]

He emphasizes the axiomatic character of Newton's work and the place of the Definitions and the Scholium to the Definitions prior to the presentation of the Axioms, the three famous laws of motion. He speaks of a modern distinction between variables as defined solely by their place in an axiomatic system, as opposed to the application of that system. The distinction between the "common" and "mathematical" notions, he asserts, is not a distinction between two kinds of things but a "logical distinction."

Next he emphasizes the importance Newton places on the example of correcting in astronomy time as measured by clocks. Unless we use an "ideal" or "corrected" time, we will have to present our theory as saying peculiar things about astronomical motion. Thus with the use of uncorrected solar time we would posit peculiar seasonal variation in the eclipsing times of the satellites of Jupiter. He draws parallels between Newton's "mathematical" concepts and the ideal points and lines (as opposed to pencil marks) of geometry. And he points out how often Newton, in saying what he means by absolute, mathematical, and philosophical time and motion, expresses himself in terms of the distinction between ideals and measures of them.

He argues that the "objective existence" of space and time was not the "central issue" for Newton, and that the bucket experiment should be read not as an empirical proof of the existence of substantival space but rather as an illustration of the need to make a "conceptual distinction" between two kinds of concepts, the common and the ideal, mathematical or philosophical. Newtonian remarks about the permanent immovability of space itself and about the equable flow of time for all time are not to be taken as the absurdities they are said to be by Mach, but rather as implicit awareness by Newton of the analyticity of assertions to the effect that some standard meets its own standard.

There is much more to the Toulmin thesis. Indeed, his main interest in the paper is in trying to throw some cold water on claims of Koyré and other that Newton's science was heavily influenced by his theological-metaphysical conceptions. The treatment of the Scholium is part of this program as Toulmin is trying to show that much in it can be reread as perfectly respectable "pure science" and not as the imposition into the science of the *Principia* of extraneous theological-metaphysical preconceptions. I will not deal with those issues here at all.

Now, it is not clear to me just what Toulmin's positive view of Newton's doctrine really amounts to, although the negative claim that Newton was not positing the existence of the "entities" space and time over and above the ordinary material entities of the world is clear. Many of Toulmin's remarks seem to rest on the old pure-versus-applied-theory distinction so familiar some years ago in the hands of those who wanted to claim that geometry was both a priori and analytic (as "pure" geometry) and a posteriori and synthetic (as "applied"geometry). But I do not think that is the most helpful way of getting at the issues here. I think, rather, that if we construe the Toulmin line as a kind of representationalist view about what Newton was saying, things become more perspicuous.

Read this way, Newton's assertion of the existence of absolute time comes down to no more than the assertion that we can assign numbers to pairs of events (their "absolute time") in such a way that the periodic processes we call clocks will generally mark off an amount of clock time more or less proportional to the number assigned the pair. Deviation of a clock from the assignment of this ideal number will be something to be explained. And the ideal time intervals we assign in this way will be the numbers to take as time differences in our equations of motion. Absolute time will be, then, in this clear way, an *idealization* of clock time. And the assertion that absolute time exists will be nothing more than the assertion that there is such a mathematical function that will assign numbers to pairs of events in just this way. The "equable flow" of time without regard to anything external will be, as Toulmin remarks, the triviality that the amount of elapsed absolute time is just the amount of elapsed absolute time, together, perhaps, with a kind of meta-theoretical assumption that when clocks deviate in the time intervals they assign from the absolute time, this deviation is to be explained in terms of some sought source of "interference" in what would otherwise be the ideal reading of the clock.

And we can construe the talk about absolute motion in the same way. We can find a function that assigns numbers to objects in such a way that the display of inertial effects by the object (the readings of an inertial guidance system strapped to it, for example) is proportional to the number assigned. It is a profound fact of nature that the difference in the numbers assigned to objects is proportional to their *relative* acceleration. But there is nothing more to say about absolute place, absolute motion, and absolute acceleration than the bald assertion that such a function with its proper connection to relative acceleration exists. Of course all of this has to be understood as what Newton was after, while making a proviso that he failed to realize that absolute acceleration was enough and that absolute place and uniform motion were uncalled for by the facts or by anything he needed to posit in this representational way of explaining them.

v

But is the representationalist reading of Newton plausible as a reconstruction of what he was after? It would seem not. And we can see this without relying upon the Queries to the *Opticks*, the letters of Clarke to Leibniz, or other extraneous evidence of a theological-metaphysical kind. In the Scholium itself Newton plainly tells us that it is by "causes and effects" that absolute motion can be known and be distinguished from motion that is merely relative. Now, one could be a representationalist and still affirm that things can be "absolutely accelerated" only when acted upon by a force—that is, by a cause. But it is hard to see how representationalism about absolute acceleration can be made compatible with the view, which Newton is plainly espousing in his bucket illustration, that it is the absolute acceleration of the material object that is *causally responsible* for the generation of the inertial effects. I think it is clear from both this illustration and from the repeated remarks about the inference to the insensible absolute motion being causal inference that Newton does intend absolute acceleration to be something "external" to the existence of the inertial effects that causally reveal it, and not merely a parameterization of the fact that in some objects these inertial features are present to a certain degree as the representationalist would read the notion of absolute acceleration.

But notice how very much better the representationalist reading of Newton sounds when it is applied to his remarks about absolute time. Here the idea of "real" time intervals as an ideal numerical assignment to pairs of events more or less accurately indicated ("measured") by more or less adequate material clocks seems to fit the words of Newton much better. And it is just in this context that talk about clock readings as a "causal effect" of the intervals of the equable flow of absolute time passing seems weirdest and most linguistically peculiar. It is for this reason that Toulmin's "modest" interpretation of Newton seems most plausible when he pulls his textual evidence from Newton on absolute time and most off the mark when he deals with the bucket experiment.

Is all of this of at most historical interest, or only of interest to someone who cares about "how things sound" rather than about how they are? Do the distinctions of the sort I have been emphasizing really just vanish away as soon as we move to a more perspicuous way of talking about space-time and its relation to the material features of the world? Perhaps; but perhaps not.

Consider the contemporary debate over realism with regard to space-time, framed as a debate between Ramsey sentence realists and Ramsey sentence representationalists. How can the issue between them be resolved? Or at least how can it be resolved if we believe contra the arguments of the pragmatists (of whom Goodman is perhaps the clearest example), who deny that there is any issue to be resolved, all ontological commitment being nothing but "talking the way some particular theory talks"?

Many suggestions have been made as to when the situation requires that we take the apparent existential commitment of a theory "seriously," as opposed to reading the existential quantifier as ranging over any abstraction that will do in a representationalist vein. Russellian-Reichenbachian-Salmonians tell us to rely on principles such as the spatiotemporal continuity of causes and the principle of the common cause, these to be accepted for either a priori or a posteriori reasons, to infer "real" hidden explanatory causes of observable effects. Whellian-Friedmanians tell us to accept the reality of hidden things when the positing of them serves to "unify" otherwise disparate observable phenomena, integrating this metaphysical posit into some version or another of confirmation as the "consiliance of inductions." Inference-to-the-best-explanationists tell us to posit as real that which is essential to provide the "best explanation" of the observable phenomena, whatever *that* means.

It is one aspect of this last general approach to theoretical realism that I want to focus on here. If one is unpersuaded that any aspect of the formal structure of theories can lead one to realism (for, after all, the formal demands are equally satisfied by the representationalist account), or that any considerations of induction or confirmation can do the trick (for, after all, any such inductive or confirmational argument can be reconstructed without loss from a representationalist perspective), then inference as resting on some notions of explanation, and associated notions of causation as something over and above regularity (and over and above regularity supplemented with spatiotemporal continuity or with "screening off" probability relations or any other merely formal device representing "unification" of the phenomena), is at least one route to try in looking for some ground for taking the putative ontological commitment of a theory as a commitment to some "concrete" reality in the world.

Certainly such an approach accords with strong intuitions on our part. Isn't the real reason we accept the reality of the hidden theoretical realm that positing it gives us our best *causal explanation* of the order we observe in the realm of the observable? And isn't it true that a mere representation of that observable order provided by the abstract interpretation of the Ramsified theory seems unsatisfactory to us because just that element of causal explanatoriness is lacking?

The problem, of course, is making some kind of coherent sense of the crucial notions invoked here. Just what would it be to "causally explain" the order in the observable, over and above embedding the description of that order in some adequate formal structure—say, the formal structure played by the Ramsified version of the theory abstractly interpreted? Our intuitions of this, for what they are worth, are, I think, intimately connected with some intuitions we have about the meanings of concepts which appear in theories at the unobservable level. Sometimes we use as theoretical terms the same terms that we use in referring to sensible objects and features of the world. Thus we speak of molecules as having size, mass, position, and so on. In other cases the theoretical concepts, "strangeness" for one example, are novel "previously meaningless" terms introduced solely in terms of their place in the theoretical network. We *feel* as though we understand the former kind of terms by knowing their meaning antecedent to the role they are playing in the theory in their new context, but that our only grasp of meaning of the latter kind is in terms of their placeholding role in the theory. Some, at least, of our intuitions about the "reality" of theoretical entities and features, and of the inadequacy of viewing their reference in the merely representational mode, and some, at least, of our intuitions that reference to the unobservables provides a *causal* explanation of the regularity among the observables and not merely an embedding representation of that regularity arise out of our intuitions about understanding what we are talking about on the level of unobservables in a manner that goes beyond the formal discourse of the theory itself.

Needless to say, all of these interconnected intuitions have been subjected to merciless criticism, and criticism difficult to overcome, by a long tradition in philosophy from Hume to Wittgenstein and beyond. I am not alleging that these ways of thinking about theoretical meaning, theoretical explanation, and realism can be made to work, but only that they are strong intuitions on our part and at least psychologically relevant to the claims that our ontological commitment in theories cannot, at least sometimes, be construed in a merely representational vein.

When it comes to Newton in the Scholium, we seem to find the notions of causal explanation more naturally fitting the relation of posited absolute acceleration and its effect in the display of the results of the inertial forces than fitting the relation of the actual "real" time intervals between events to the intervals as recorded on material clocks. Why this is so is not very clear to me, but it is certain that Newton finds himself most comfortable in talking about unobservables known by their effects when dealing with absolute motion rather than absolute time, and I think most of us would share his intuitive reaction. Not surprisingly, then, when it comes to interpreting Newton, the representationalist reconstruction of his thought seems most plausible when we read him dealing with absolute time intervals and their "sensible measures" and least plausible when we try to reconstruct his intentions in his talk about absolute motion and the origin of inertial forces.

Beyond the problem of the historical interpretation of Newton, there is the issue of the current debate between the "realist" understanding of space-time theories and relationists of various modern stripes. I think that at least part of the reason why space-time theories seem to be taken as such a crucial test case for theoretical realism versus irrealism or representationalism is that space-time is such a peculiar entity and its relation to the "sensible" such a peculiar relation.

When we reflect on the water sloshing in the spinning bucket, the vertigo we feel in the amusement park ride, or the enormous stresses necessary to disintegrate the overrevved flywheel made of high-strength steel, we cannot believe that it isn't something real—say, real timelike geodesics of a real space-time and real deviation of the trajectory of the test object from such geodesics—that is *causally* responsible for the force, much as the real relative velocity of charges is responsible for their magnetic interaction. We are not satisfied with relative acceleration as the sufficient explanation of the difference in the inertial forces felt by two test objects, but we want to know why it is this one and not that one that feels the forces experienced. Explanation as causation and analogy with relative motion and its causal efficacy lead us to feel the pull of the "realist" interpretation of space-time theories.

But when we think about the variety of material clocks ticking away more or less uniformly relative to one another, the idea that there is a way of assigning numbers to pairs of events such that each such clock more or less accurately reveals that number seems to be the core of talking about time itself. Of course, deviation of any clock from its ideal rate is something to be explained by causal interaction in the material world. But there is no "causal" explanation as to why clocks in general record time intervals more or less accurately. What we mean by time intervals is just this numerical abstraction and idealization from the uniformity more or less of relative rates of clocks of various kinds of construction. It is, of course, still an important observation of Newton's that only when we date events by the ideal time metric will our dynamical laws of nature take on their familiar simple form. But that does not seem to call for absolute time as a "cause" either. There are those who are happy to say that the elapsed proper time along a timelike path somehow "causally" explains the time reading of, say, an atomic maser transported along that path. But for most of us that sounds like confusing a mere ideal with a concrete cause.

Now, in the fullness of coherent philosophy none of this may survive. Indeed, we may ultimately agree with the pragmatist that the very distinction between a representationalist and a realist reading of theory is a mythical one. Whatever our ultimate conclusion, though, I think we are at least psychologically inclined initially to draw distinctions of the kind I have been noting here. And I think that these distinctions go some way toward explaining our sense that Newton's language becomes peculiar when applied uniformly across the phenomena he discusses in the Scholium, in explaining how such contrary readings of Newton's intentions can be supported by those such as Mach who take him to be a realist and those such as Toulmin who read him as a representationalist, and in explaining at least part of the intuitive impetus behind some of the arguments brought to bear both in favor of realism and against it and in favor of representationalism by contemporary interpreters of space-time theories.

Notes

1. I. Newton, "Scholium to the Definitions," in *Mathematical Principles of Natural Philosophy*, trans. Motte, rev. Cajori (University of California Press, 1947), p. 11.

2. Ibid., pp. 6–12.

3. "Axioms, or Laws of Motion," ibid., pp. 20–21.

4. S. Toulmin, "Criticism in the History of Science: Newton on Absolute Space, Time, and Motion, I," *Philosophical Review* 68 (1959), 1–29, and "Criticism in the History of Science: Newton on Absolute Space, Time, and Motion, II," *Philosophical Review* 68 (1959), 203–227.

Chapter 5

Absolute Time versus Absolute Motion: Comments on Lawrence Sklar

Phillip Bricker

i

Three hundred years after the publication of Newton's Scholium on absolute space, time, and motion, the debate between absolutists and relationalists is as vigorous as ever. Earlier in this century it was widely held that Einstein's theory of relativity showed once and for all the falsity—if not the incoherence—of Newton's absolutist position. But much progress has been made since then. We now know that Einstein and other modern relationalists tended to conflate different senses of "absolute." Relativity theory teaches us that space, time, and motion are not absolute only for *some* senses of "absolute." Once the different senses have been carefully distinguished, it becomes clear that philosophically important versions of absolutism can survive the advent of relativity theory (although absolute space and absolute time must be traded in for absolute *space-time*).[1]

It is especially important to distinguish those senses of "absolute" that do and those that do not have ontological implications. *Ontological absolutism*—or *realism*—holds that space and time[2] are entities that exist over and above the objects and events of the material world. Reference to space and time is to be take literally. When one asserts that a material object occupies a part of space or endures through an interval of time, one asserts that a genuine relation holds between distinct entities, one material and one immaterial, not merely that some complex property applies to material things. Indeed part—or the whole—of space or time could exist even though no matter occupied that space or endured through that time. *Ontological relationalism*—or *representationalism*—by contrast, holds that space and time exist only as mathematical models or representations of the spatiotemporal properties and relations among material objects and events; if there were no matter, there would be no space or time. On this view, all reference to space and time is to be interpreted by way of complex structural properties of the material world. It is important to emphasize that the debate between realists and representationalists is over whether space and time have independent existence, not over whether spatiotemporal properties and relations are in some sense reducible to observable features of the world. The representationalist, as well as the realist, can allow unobservable spatiotemporal properties and relations among material objects and events.[3]

In "Real Quantities and Their Sensible Measures," Larry Sklar tries to uncover the competing intuitions that have fueled the realist-representationalist debate. He returns to Newton's Scholium, and claims that two rather different sorts of problem are being addressed. The *problem of absolute time* is concerned with the question of which material clocks, if any, provide an accurate measure of time, and how inaccurate clocks may be corrected. The *problem of absolute motion* is concerned with the question of which material objects are truly at rest, which in motion, and how the quantity and direction of such motion—and thus the object's space-time trajectory—can be determined. This latter problem may be divided (although Sklar does not do so) into two subproblems: the *problem of absolute velocity* and the *problem of absolute acceleration*. These problems are related as follows. Information about the absolute velocity of any object throughout an interval of time (magnitude *and direction*) completely determines[4] the object's space-time trajectory, and so provides all there is to know about its properties of absolute motion throughout that interval; information about the absolute acceleration of the object constrains, but does not completely determine, the object's space-time trajectory.[5] Newton notoriously failed to distinguish explicitly the problems of absolute velocity and absolute acceleration in the Scholium, but it will behoove us to do so in what follows.

Why does Sklar think it important to separate the problems of absolute time and absolute motion? There has been a tendency on both sides, he claims, to assume that arguments for or against realism with respect to one problem would apply *mutatis mutandis* to the other. But this, he thinks, may not be so. In particular, the realist arguments seem strongest when applied to the case of absolute motion—specifically absolute acceleration—but sound wrong or peculiar when applied to the case of absolute time; the representationalist arguments stand in the reverse relation. The reason for this, Sklar suggests, is that the inertial effects of absolute acceleration cry out for causal explanation, and such explanation, to be satisfying, will tend to invoke the reality of space and time, whereas the distinction between accurate and inaccurate clocks can be understood without invoking an immaterial entity, such as absolute time, as a causal agent.

In what follows I will argue that the distinctions Sklar draws do not go very deep: they depend on the fact that absolute acceleration is, by definition, a property of material objects, whereas absolute time is not. If the problems of absolute time and absolute acceleration are characterized in a parallel fashion in terms of the temporal and motional properties of material objects and events, the distinctions Sklar draws tend to vanish. The realist can and should apply his arguments and intuitions equally to both problems; the representationalist can and should do the same. I thus conclude that Sklar's distinctions do not go to the heart of the realist-representationalist debate.

ii ·

I would like to begin by discussing Sklar's interpretation of Newton's Scholium.[6] According to Sklar, the tendency to treat the problems of absolute time and absolute motion alike, and the resulting peculiarities of language, originate with Newton himself. For example, by applying to the case of motion talk about sensible measures that is appropriate to the case of time, Newton is led to the peculiar assertion that relative motion provides a sensible measure of absolute motion. Conversely, by applying to the case of time talk about causes and effects that is appropriate to the case of motion, Newton (or his expounder) is led to peculiar assertions such as that the intervals of absolute time cause material clocks to tick the way they do. Although Newton may try to avoid saying some of these things, he is committed to them nonetheless. Or so Sklar claims.

I will argue that the first peculiarity having to do with sensible measures can be understood without supposing that Newton conflated the problems of absolute time and absolute motion. My main focus, however, will be the second sort of peculiarity having to do with causation. I will argue that it is not to be found in anything Newton said or would be led to say, and that Sklar's claim that Newton treats absolute time and absolute motion alike with respect to talk of causes and effects is ungrounded.

It is odd on the face of it to say that Newton treats alike the problems of absolute time and absolute motion. The Scholium begins by separately characterizing and illustrating, for each of the quantities time, space, place, and motion, the distinction between absolute and relative. It then contains a paragraph devoted exclusively to the problem of absolute time, the problem of how absolute time can be distinguished from relative time, and why such a distinction is necessary. It concludes with a lengthy discussion devoted exclusively to the problem of absolute motion with reference to the renowned rotating bucket and the revolving globes. Indeed the only place Newton treats absolute time and absolute motion together is in the summarizing paragraph quoted by Sklar at the beginning of his paper (and in the Scholium's introductory remarks). However, in that paragraph, it is true, the relative quantities are all treated alike in at least one respect—namely, that they are all sensible measures, accurate or inaccurate, of the real quantities, from which it follows in particular that relative motion is a sensible measure, accurate or inaccurate, of absolute motion.[7]

Why does this claim sound peculiar? Newton could have said: relative motion is some measure of motion relative to a body assumed to be at rest; it is an accurate measure of absolute motion when the reference body is at rest in absolute space, otherwise inaccurate. Given Newton's assumptions, this definition of when the measure is accurate or inaccurate is perfectly meaningful. The problem, I take it, is that it is not possible, even in principle, to determine whether the measure is accurate or inaccurate because it is not possible to determine whether the reference body is at rest in absolute space. In Newtonian mechanics one can measure the absolute *acceleration* of the body by measuring the forces acting on it;[8] but one cannot infer the absolute *velocity* of the body, which might have any value whatsoever. Thus the accuracy of the measure cannot, even in principle, be established.

On this account Newton's statement is peculiar because the relation between relative motion and absolute motion is too weak. If a sensible measure is to count as a measure *of* some quantity, there must be an appropriate connection between the sensible measure and the quantity to be measured, some lawlike or causal connection that allows one to infer the real quantity from the observations of measurement (perhaps together with other observations). For example, within Newtonian mechanics the solar day provides a sensible measure, albeit inaccurate, of absolute time because the laws of celestial mechanics determine the length of the solar day as a function of time, which then allows the true elapsed time to be calculated from the sun's apparent location by way of the inverse function. Relative motion, by contrast, does not provide a sensible measure of absolute motion within Newtonian mechanics, accurate or inaccurate, because there is no lawlike connection that would allow one to infer the quantity of absolute motion from observations of relative motion, even together with observations of forces.

Might not, however, the notion of being a sensible measure of a quantity be susceptible of a weaker interpretation, one that allows a weaker connection between the measure and the measured quantity?

On this interpretation the measure must provide *some* information about the quantity being measured, or a component of the quantity, but it need not be possible to infer everything about the quantity from the measurement data. I want to suggest that Newton may simply have had this weaker interpretation in mind. This is confirmed by Newton's final illustration: the case of the two globes surrounded by bodies whose position is fixed relative to one another (p. 12). Newton uses this case to illustrate how relative motion can provide a measure of absolute motion. He argues that the state of absolute circular motion of the fixed bodies—both its quantity and its direction—can be determined by measuring the bodies' circular motion relative to the two globes, whose state of absolute circular motion has been determined by measuring the tension in the connecting cord. Newton clearly considers this a case of measuring absolute motion, even though it is only absolute *circular* motion that is measured—that is, absolute acceleration but not absolute velocity. This suggests that Newton has the weaker interpretation of "sensible measure" in mind. Perhaps his claim would sound less peculiar if explicitly qualified: relative motion (in conjuction with forces) provides a sensible measure of *some* kinds of absolute motion—namely, circular motion or accelerated motion.[9] In any case, the peculiarity in Newton's language, if there is any, would not result, as Sklar suggests, from a conflation of the problems of absolute time and absolute motion.

iii

Let us turn now to the second sort of peculiarity having to do with causal attributions. According to Sklar, Newton treats absolute time and absolute motion alike with respect to questions of causation: in both cases, they are inferred by some sort of causal inference from observable phenomena, and thus both absolute time and absolute motion are, in some sense, causes that have material effects. This dual treatment will compel Newton to say not only sensible things, such as that absolute motion causes inertial forces, but also peculiar things, such as that the intervals of absolute time cause clocks to tick the way they do.

But on what grounds does Sklar claim that Newton treats absolute time and absolute motion alike with respect to causation? The discussion of causes and effects in the Scholium is devoted entirely to absolute motion; one searches in vain for any passage in which Newton speaks of the causes or effects of absolute time. Yet, according to Sklar, Newton holds across the board that "the real quantities are known to us only by their causes and effects, that is, by causal inference from the behavior of the sensible measures which are available to our sensory inspection." In fact, Newton never talks about the causes and effects of any real quantity other than absolute motion.

With respect to absolute space, Newton explicitly asserts that it cannot be known by its causes and effects. He twice emphasizes in the Scholium that the parts of absolute space "by no means come under the observation of our senses" (p. 12).[10] I thus find it doubtful that any of Newton's arguments in the Scholium can be interpreted as involving a "causal inference" to the existence of absolute space, as Sklar seems to think. Newton's inference to absolute space appears to be much less direct, perhaps by way of the assumption that absolute motion entails absolute space, perhaps by way of more general theoretical considerations of the sort Sklar mentions toward the end of his paper. In any case, realism about space and time is presupposed in the Scholium, not argued for directly.[11] Newton's arguments in the Scholium are primarily concerned with questions of reducibility and observational distinguishability. Can an object's absolute spatiotemporal properties be defined in terms of its sensible relations to other objects? If not, how can its absolute spatiotemporal properties be distinguished observationally from those properties that are merely relative? As I mentioned at the beginning of this paper, these questions are independent of the ontological debate; or at least the connection requires argument and cannot be taken for granted.

Thus I cannot agree with Sklar's claim that, with respect to causation, Newton treats all the real quantities, and in particular absolute time and absolute motion, alike. Indeed the basis on which Newton treats the various quantities differently with respect to causation is not, I think, far to seek. It has to do with a fundamental difference in the way these quantities are initially characterized. Absolute space and absolute time are entities that exist independently of material objects and events; not so for absolute motion. Absolute motion is defined by Newton as "the translation *of a body* from one absolute place to another" (p. 7, my emphasis). It is thus a *property* or *state* of material objects; if there were no material objects, there would be no motion.[12] Thus, the fact that Newton speaks about the causes and effects of absolute motion, but not of absolute space and time, can be understood on the following simple hypothesis: Material objects, their properties and states stand in causal relations; space and time, which are immaterial entities, do not. What makes Sklar's causal statements about absolute time peculiar is that they violate this natural precept. Newton would certainly reject them.

iv

There is a sense, however, in which Newton can and would treat absolute time and absolute motion alike with respect to causation. This sense emerges if, in comparing causal statements about absolute motion with causal statements about absolute time, we make sure to compare like with like—something Sklar fails to do. As I have just noted, Newton's statements about the effects of absolute motion are always about the absolute motion *of some material object or system*; for example, the absolute motion *of the water* causes the surface to go concave. If Newton were to treat absolute time analogously to absolute motion in this respect, this would require him to attribute material effects not to the intervals of absolute time itself (as Sklar would have it), but rather to the duration of material events and processes; and causal statements of this sort are by no means peculiar. For example, the telephone's ringing for thirty seconds caused my answering machine to pick up; or, my watch's running slow caused me to miss the meeting. It is statements of this sort that are analogous to the causal statements Newton cites about absolute motion, not statements claiming that the intervals of absolute time cause material clocks to tick the way they do. In sum, we can say that material objects or events have causal powers by virtue of their motional *and temporal* properties without having to say that absolute time itself has material effects.

But is the case of temporal properties fully analogous to the case of states of motion? Newton argues not just that the absolute motion of objects has material effects, but that those effects can be used to distinguish an absolute from a purely relative motion. Can we also say that the absolute duration of events has effects that allow us to distinguish absolute from relative time? Indeed we can. For Newtonian mechanics allows us to distinguish by the measure of centrifugal forces not only the state of absolute rotation from absolute nonrotation but also the state of absolute *uniform* rotation from absolute nonuniform rotation. Now, suppose we have an object in some state of absolute rotation. Any such object is a clock that ticks off one unit of time for each complete rotation. Whether or not such a clock keeps absolute time can be determined by examining the causes and effects of the ticking of the clock, just as whether a body is in absolute rotation can be determined by its causes and effects. Thus, absolute time and absolute motion can be treated alike with respect to causation; both the absolute uniform ticking of a clock and the absolute motion of an object can, at least sometimes, be determined by their causes and effects.

Absolute space, however, is different. The problem of absolute space, for Newton, is the problem of determining whether an object is in the same absolute place at two different times. Since this property of being in the same absolute place at different times has no material effects within Newtonian mechanics, it is not possible to distinguish absolute from relative space by examining causes and effects. This suggests that, at a deeper level, the problems that should be kept distinct are not, as Sklar claims, those of absolute time and absolute motion, but those of absolute time and absolute motion on the one hand, and that of absolute space on the other. Or, to be more exact, the problems of absolute time and absolute *acceleration* on the one hand, and absolute space and absolute *velocity* on the other. Properties of material objects and events involving absolute space and velocity are unobservable by any means; properties involving absolute time and acceleration are observable, at least indirectly, by means of forces and material clocks. This might explain why Newton twice emphasizes that absolute space is insensible and in no way comes under our observation, whereas he never asserts the same about absolute time or absolute motion. Am I here disagreeing with Sklar's claim that, according to Newton, *none* of the real quantities is "available to the senses"? Perhaps by "not available to the senses" Sklar means only "cannot be sensed without the mediation of material objects or events," in which case there is no disagreement. I am claiming that Newton may have meant something stronger than this when he asserted the insensibility of absolute space, something that does not apply across the board to all the absolute quantities.

Let us return to the peculiar causal statements about absolute time. I have argued that one can treat absolute time and absolute motion (that is, acceleration) alike with respect to causation, in both cases focusing on properties of material objects or events, without being committed to the peculiar statements in question. But what if we focus in both cases on the immaterial entities—space, time, or space-time? I claim that we can still treat absolute time and absolute motion alike with respect to causation. For the case of motion, we should ask about the effects of the four-dimensional affine structure of space-time, since it is the affine structure that determines which space-time trajectories are and are not absolutely accelerated. What corresponds to the peculiar statement that the intervals of absolute time cause clocks to tick uniformly is then not that the absolute motion of some material object causes certain inertial effects, but rather that the affine structure of absolute space-time causes material objects to move the way they do. And that latter statement sounds just as peculiar as the statement about intervals of absolute time. Newton can hold that absolute space, time, and space-time play a role in causal explanations without taking them to be *causes* of material events. Thus, he might say that material objects and events have causal powers in virtue of occupying the regions of absolute space-time that they do. But causal powers need not ever be attributed to the regions of absolute space-time themselves.

I say "need not." In Einstein's general relativity absolute space-time becomes a dynamical object, and it becomes more natural to attribute causal powers to space-time itself. In general relativity it is common to say things like: Matter causes space to curve, which in turn causes matter to move as it does. But in Newtonian theory there can be no genuine interaction between matter and the structure of space because the structure of space is immutable; and that, I think, is why it sounds peculiar to attribute causal powers to absolute space itself. Thus we *need not* attribute causal powers to absolute space and time, and, within the context of Newtonian theory, if we do not want to sound peculiar, we *should not*. Moreover, if what I have said has been correct, Newton *did not* and *would not* attribute such powers to absolute space and time themselves.

v

Sklar's purpose in separating the problems of absolute time and absolute motion is not simply historical. He thinks it helps to shed light on the current debate as well. The case for realism is strongest when applied to the problem of absolute motion: "When we reflect on the water sloshing in the spinning bucket . . . we cannot believe that it isn't something real, say real timelike geodesics of a real space-time . . . which is *causally* responsible for the force." The problem of absolute time, however, supports representationalism, not realism: "But when we think about the variety of material clocks ticking away more or less uniformly relative to one another, the idea that there is a way of assigning numbers to pairs of events such that each such clock more or less accurately reveals that number seems to be the core of talking about time itself . . . There is no 'causal' explanation as to why clocks in general record time intervals more or less accurately." Sklar is reporting intuitions, not subjecting them to analysis. But I have doubts whether, even at the level of intuition, Sklar's contrast holds up under scrutiny. I will argue that the "realist" intuitions fostered by contemplating absolute motion are compatible with representationalism, and the "representationalist" intuitions fostered by contemplating absolute time are compatible with realism.

First, let us consider the problem of absolute motion. Let us grant

Sklar's assumption that the inertial forces generated by rotating buckets and the like demand *causal* explanation. When there is a "difference in the inertial forces felt by two test objects, [we] want to know why it is this one and not that one that feels the forces." Moreover, this requires positing some *cause* that makes the difference. Still, a causal explanation of the inertial forces might take any of three quite different forms:

(1) The inertial forces are caused by properties or states of motion of the objects in question, such as absolute acceleration. These properties are primitive, and do not entail the existence of immaterial entities: space, time, or space-time.

(2) The inertial forces are again caused by properties or states of motion of the objects in question, but the properties are not primitive. Objects have these properties by virtue of their relations to immaterial entities: space, time, or space-time. The immaterial entities are not themselves causes of the inertial forces, or of anything else.

(3) The inertial forces are directly caused by the action of an immaterial entity—space, time, or space-time—upon the material objects in question.

A representationalist who accepts the demand for causal explanation will offer an explanation of type (1); a realist will offer one of either type (2) or type (3). The difference between (2) and (3), I suspect, is mostly verbal; as I have noted, (3) sounds peculiar unless space-time is taken to be a dynamical object, but the ontological commitments will be the same either way. The realist rejects (1) on the grounds that positing primitive properties of absolute acceleration without positing absolute space and time, or space-time, is ad hoc, or unintelligible, or in some way theoretically unsatisfactory. The representationalist rejects (2) and (3) on the grounds that positing an immaterial space and time, or space-time, is otiose, or unintelligible, or in some other way objectionable. It is over grounds such as these that the battle between realists and representationalists will be fought. But both sides, it seems to me, can agree that rotating buckets and the like call for causal explanation. Why, then, does Sklar claim that the case of absolute motion supports realism over representationalism?

Sklar must think that a causal explanation of type (1) is for some reason ruled out for the representationalist. Perhaps he does not think the representationalist can help himself to a *primitive* property of absolute acceleration, and so lacks any means to explain the inertial forces. Thus Sklar writes at an earlier point: "It is hard to see how representationalism about absolute acceleration can be made compatible with the view . . . that it is the absolute acceleration of the material object which is *causally responsible* for the generation of the inertial effects." But the representationalist, as far as I can see, need not reject absolute acceleration as a property possessed by material objects *over and above* the felt inertial forces: the representationalist *reinterprets* absolute acceleration in a way that avoids any commitment to a separate entity, space-time; he need not *eliminate* it in favor of the inertial forces. Thus, he is free to causally explain the inertial forces along the lines of (1). At any rate, it requires realist argumentation to see why such explanation is unsatisfactory, not merely the "realist" intuitions put forth by Sklar.[13]

What about the problem of absolute time? Sklar claims that "there is no 'causal' explanation as to why clocks in general record time intervals more or less accurately." But what can this mean? Certainly we would want to explain causally why some clock is inaccurate just as much as why some water is sloshing in a rotating bucket. For any mechanical clock, its accuracy or inaccuracy will be explained at least in part in terms of the inertial forces present, which in turn require the very same sort of causal explanation required in the case of absolute motion. So why should contemplating the problem of absolute motion lead to realism any more than contemplating the problem of absolute time?

Perhaps when Sklar says that the behavior of clocks does not require "causal" explanation, he means only that it does not require causal explanation of type (3), explanation that invokes absolute space, time, or space-time as a cause. This is confirmed by the fact that he uses "call for causal explanation" and "call for absolute time as a cause" interchangeably in the section under discussion. But both the realist and the representationalist can agree in rejecting explanations of type (3). Thus, the fact that the problem of absolute time does not call for explanations of this type cannot support representationalism over realism.

I suspect what has happened is that Sklar has ignored explanations of type (2), thus creating a false dilemma. He assumes here—as he did with Newton—that the realist will support his case by appealing to some sort of direct causal inference to the existence of space, time, or space-time. Then, since such an inference to absolute time as a cause of material events is especially implausible, he concludes that the case for realism is weak when applied to the problem of absolute time. But if the realist's inference to space, time, or space-time is less direct, if his causal explanations are of type (2), then he need not respond to the problem of absolute time as Sklar suggests. The realist need not choose between a representationalist view of time and the view that absolute time somehow causes clocks to behave the way they do. Sklar has posed a false dilemma.

In conclusion, the realist can consistently apply his intuitions to both the problem of absolute motion and the problem of absolute time; the representationalist can consistently do the same. For this reason I do not think the distinctions Sklar focuses upon go to the heart of the debate between realism and representationalism.

Notes

1. For an untangling of some of the different senses of "absolute," and a recent defense of absolutism, see Michael Friedman, *Foundations of Space-Time Theories* (Princeton: Princeton University Press, 1983).
2. Or space-time. But for convenience I speak prerelativistically in what follows, unless otherwise noted.
3. See Friedman, op. cit. pp. 217–223, for a more precise characterization of the ontological debate in terms of the models of space-time theories.
4. Up to isometries of space and of time.
5. If one focuses instead on instantaneous states of motion, the relation changes: neither absolute instantaneous velocity nor absolute instantaneous acceleration suffices to determine the object's complete state of absolute motion at a time; higher-order derivatives are needed as well.
6. The Scholium on space, time, and motion occurs in the *Principia* just after the initial definitions and before the Axioms, or three laws of motion. Page references will be to Sir Isaac Newton, *Mathematical Principles of Natural Philosophy and His System of the World*, trans. Andrew Motte, rev. Florian Cajori (Berkeley: University of California Press, 1934).
7. Note that "more or less accurate" is Sklar's addition; Newton says "accurate or inaccurate" (p. 11).
8. Assuming, at any rate, that one can exclude the possibility of external forces, an assumption Newton freely makes in the Scholium.
9. Related omissions of explicit qualifiers occur elsewhere in the Scholium. For example, Newton writes: "We may distinguish rest and motion, absolute and relative, one from the other by their properties, causes, and effects" (p. 8). The force of this statement must be only that we may *in some cases* make the distinction, as Newton is well aware.
10. In his earlier unpublished work, "De Gravitatione et Aequipondio Fluidorum," Newton uses the causal inefficacy of absolute space as one of the chief features that distinguishes it from material substances (the other being immovability). Space is not "an entity that can act upon things" and is not capable of "exciting in the mind sensation or perception." (I owe this reference to Howard Stein.) See A. Rupert Hall and Marie Boas Hall, eds., *Unpublished Scientific Papers of Isaac Newton* (Cambridge: Cambridge University Press, 1962), pp. 121–156.
11. This view is cogently argued in Ronald Laymon, "Newton's Bucket Experiment," *Journal of the History of Philosophy* 16 (1978), 399–413. He summarizes: "It is not intended by Newton that these experiments [the rotating bucket and the two globes] have as their conclusions the existence of absolute space, since this existence is already assumed by their explanation. The only support that these experiments give for the existence of absolute space is that they show that this concept

does have some application and is part of a successful scientific theory'' (pp. 410–411).

12. In this connection, note that prior to the Scholium Newton defines *quantity of motion* as mass times velocity—what we call momentum.

13. In *Space, Time, and Spacetime* (Berkeley: University of California Press, 1974), Sklar considers a view that posits a property of absolute acceleration without positing absolute space and time, or space-time. He denies that such a property can be used to *causally explain* the inertial forces, but no argument for this is given. See pp. 229–234.

Chapter 6

Predicates of Pure Existence: Newton on God's Space and Time

J. E. McGuire

Some years ago I argued that Newton's doctrine of absolute space and time is motivated by his view of the existence of a divine being.[1] In the course of this study I advance the opinion that Newton attempts "to distance" space and time from divine essence. In effect, I argue that Newton links the infinity of space and time to divine existence, which he then associates with the actuality of God and not directly with his essence. Here I wish to return to this view in the light of further thinking and in response to some of John Carriero's observations on my earlier views. Also, I want to reconsider my claim that Newton's conception of how space and time relate to God's existence is not an instance of causal dependence. I think now with Carriero that the dependence may be taken as causal, but, if this is so, the distinction between ontic (my earlier view) and causal dependence is extremely attenuated indeed. If I spell out my agreements and disagreements with Carriero's commentary, I do so not to have the last word but to advance the dialogue.

The first part of this study deals with the nature of Newton's religious sentiment and its associated theological framework. The second considers the metaphysical position that these views engender and its implication.

i. God and Worship: The Theology of a Living God

An obvious fact about Newton is the pious and biblical nature of his religious experience. To follow the Christian dispensation is to give unwavering obedience to the commandments of God. Moreover, it is our duty to seek knowledge of God from the evidence of his works: in this way we give him honor for the glory and design of creation. It is true, as Manuel claims, that Newton distrusts the conceits of abstract metaphysical theology.[2] It is wrong, however, to suppose that he avoids altogether the theological implications of his religious views. In a manuscript of the early 1690s, which I entitle "Tempus et Locus," this "theological turn" is made abundantly clear.[3] The manuscript indicates Newton's particular theological approach to questions concerning the religious experience of God. It shows, moreover, that a close relationship obtains for Newton between his view of the existence of God and his conception of absolute space and time. Furthermore, the manuscript provides understanding of the motives that inform Newton's conception of God's existence. In turn, these motives help to clarify the implications of his conception of divine presence in relation to his views on the nature of space, time, and creation.

Even a casual reading of the manuscript indicates that Newton conceives God as a real person and not as an abstract metaphysical being. God is a living, intelligent, and powerful agent who always and everywhere exists. He is likened to an absolute king who freely decrees the law to all created things. To think of God as existing beyond space and time is an unduly abstract conception. It is not easily comprehended by the mind and fails to promote suitable attitudes of worship (pp. 121–123). In the first draft of the manuscript Newton in fact characterizes abstract conceptions of God as arising "ex Scholasticorum disputationibus." The reference is of course meant to be pejorative. And in fact the metaphysical views of God found in the writings of many Scholastics do indeed differ from Newton's. Newton's God is a biblical God of dominion whose immediate presence in nature purposively enacts the destiny of all created things. Thus providence is directly grounded in divine uniquity and omnipresence. And God, in virtue of his actual presence in space and time, is "able to act in all times & places for creating and governing the Universe."[4]

In developing the details of his theological position in "Tempus et Locus," Newton rhetorically juxaposes eight contrasting opinions concerning divine nature. In each case it is the second opinion that he supports, and for which he urges acceptance. They constitute an interesting set and go beyond anything he had previously written.

First, Newton rejects the view that God's existence is "all at once," as expressed by the phrase *totum simul*. That is, he affirms that God's existence can be characterized by successiveness and the temporality of earlier and later. Indeed he calls God's life by the name "He that was and is and is to come" (p. 121). Thus in Newton's conception God's existence is sempiternal; for there is no time past, present, or future at which God does not exist. This view is also expressed in the *Principia* (1687): "Since every particle of space is *always*, and every indivisible moment of duration is *everywhere*, certainly the Maker and Lord of all things cannot be *never* and *nowhere*."[5] Accordingly, God does not exist in a timeless present; he endures in unending time, his living existence devoid of beginning and end.

That Newton does not conceive God's existence as timeless is supported by two further passages. In a draft variant to the General Scholium he states: "His duration is not a *nunc stans* without duration, nor is his presence nowhere." He writes to Des Maizeaux in a similar vein in 1717: "The Schoolmen made a *Nunc stans* to be eternity & by consequence an attribute of God & eternal duration hath a better title to that name, though it be but a mode of his existence. For a *nunc stans* is a moment wch always is & yet never was nor will be; which is a contradiction in terms."[6] Thus, for Newton, God's living existence is irreducibly durational. To think of him as existing in a *nunc stans*—a stationary and timeless now—is to deny that he exists. For Newton the actuality of anything that exists must be successive and involve time: "What is never and nowhere, it is not in *rerum natura*" ("Tempus et Locus," p. 117). Furthermore, the conception of a "stationary and timeless now" is for Newton contradictory. What would it be like, Newton seems to ask, to exist in a timeless moment? Can anything be said actually to exist in an unchanging and indivisible moment? Do we have a coherent conception of an unchanging "now" which always is? For Newton this sort of talk is parasitic on the tense structure of language. It has significance only if it is contrasted with what was and what will be. Thus, those who use the phrase *nunc stans* to refer to the mode of God's existence, are covertly employing tensed forms of *is* while pretending that tensed expressions are inappropriate to the manner of divine being. According to Newton, then, although the phrase is meant to capture a timeless mode of existence, it in fact implies temporal devices. For Newton it is best to conceive God's existence as sempiternal, since the mind can grasp this notion. Moreover, it is a coherent notion, and accords better with the biblical conception of divine existence, where divine presence is not unambiguously debarred from temporality. An apologist is hard-pressed (as Newton well knew) to find biblical passages that straightforwardly support a conception of God's existence in terms of the phrases *totum simul* and *nunc stans*.

The conception of eternity that Newton rejects has its first systematic articulation in Plotinus' *Ennead* III.7.[7] In the West, the doctrine of eternity as the eternal "now," which derives from Plotinus' treatise on eternity and time, appears in the writings of Augustine and Boethius, and through them the *nunc stans* and *totum simul* of eternity pass into medieval and subsequent thought. To any thinker of Plotinus' sensibility the notion of an eternal life apart from duration and change is neither self-contradictory nor incoherent, nor does the use of temporal language imply any duration or temporalization of eternity. Plotinus himself makes use of Aristotle's conception of ἐνεργεία in articulating his notion of the durationess and eternal life of νοῦς. At *Metaphysics* θ. 1048b. 21–23, Aristotle distinguishes between two sorts of actuality (which he calls κίνησις and πρᾶξις): the first sort *has* an end or τέλος, while the second sort *is* an end. Seeing, for example, is an activity of the latter sort. As such, it cannot be analyzed into stages leading to its actualization, since it is by its nature complete at every point. Hence, in a complete activity like seeing, which is itself an end, there is no distinction between coming to activate a potential for seeing and the completion of this activation. Hence no duration need be involved in such an activity: its actualization is instantaneous. This Aristotelian notion of ἐνεργεία is clearly the model for Plotinus' conception of the eternal life of νοῦς, and he seems right to suggest that there need be nothing intrinsically durational or temporal about it.

This Plotinian view of the eternal ζωή is deeply antithetical to Newton's anthropomorphic conception of divine nature. In Newton's view a deity worthy of human worship must exist in the world actually, substantially, and intimately. God must therefore be able to exist at all times and in all places and possess an intelligent life that literally exists through unending duration. For Newton this view cannot be captured by the atemporal notion of duration implied by the *nunc stans* doctrine, nor, of course, by the view that God exists supratemporally. In Newton's eyes his conception alone provides proper motivation for Christian belief and worship, and coheres with the biblical view of God as an individual person who is able to act intentionally and purposively in nature, and providentially through history. In the second section I return to Newton's anthropomorphic account of God, in particular to his attempt to envision God according to the same composition of essence and existence that holds of finite things. There is a clear connection between this view and Newton's conception of how space and time relate to divine existence.

Newton also believes that God exists literally in infinite space. This means two things. Considered from the perspective of his living existence, God is said to exist necessarily in every place at all times whatsoever: but considered by virtue of his individual nature, God's immensity and omnipresence have reference to space itself. Of God's existence in all places whatsoever, Newton says in his Twelve Articles of faith: "The Father is immovable, no place being capable of becoming emptier or fuller of him than it is by the eternal necessity of nature. All other beings are movable from place to place."[8] Essentially the same is said in the Yahuda manuscript: "For God is alike in all places, He is substantially omnipresent, and as much present in the lowest Hell as in the highest heaven."[9] God's existence is thus without limitation in every place of space, for space itself is in fact infinite.

But God does not merely exist sempiternally in infinite space. According to Newton, it is appropriate to his supreme perfections that he is always active. Thus, God can and does act in space and through time. In a draft fragment to the first version of the "Tempus et Locus" manuscript Newton asks approvingly "whether the Prophets more correctly say that God is present absolutely in all places, and constantly sets in motion the bodies contained in them acording to mathematical laws, except where it is to the good to violate those laws."[10] A related view appears in a late draft version of the scholium on space and time of the *Principia* (1687). After remarking (in a passage that later appears in print) on the difficulty of distinguishing true from apparent motions with respect to the parts of absolute space, Newton begins but does not finish the following sentence: "Solus enim Deus, qui singulis immobiliter et insensibiliter." Following Cohen's interpretation, and the context of the passage, the sentence may reasonably be completed as follows: "For God alone, who [gives motion to] individual [bodies] without moving and without being perceived, [can truly distinguish true motions from apparent]."[11] Thus, not only does God exist in infinite space, and not only is he immediately present in all created things, but he can act directly on things that exist in the vast receptacle of space; for "all spaces are and always have been equally capacious of containing things" ("Tempus et Locus," p. 121). In Newton's view, God's creative power is clearly unrestricted in scope. He is not limited to creating the present world, so long as logical impossibilities are not in question. For God cannot do what is impossible (undo what is done), any more than he can do anything that implies an imperfection (for example, lie or deceive).

Given these commitments, it is not surprising that Newton claims God "is omnipresent not *virtually* only, but also *substantially;* for power cannot subsist without substance. In him are all things contained and moved; yet neither affects the other: God suffers nothing from the motion of bodies; bodies find no resistance from the omnipresence of God."[12] This is a characteristic expression of Newton's conception of divine omnipresence. In "Tempus et Locus" he states unambiguously that the infinity of God's space is the expression of his "eternal omnipresence" (p. 121). Because he actually and substantially exists in infinite space, God can act in and at every place. The Cartesian conception of a God who is everywhere according to power, but nowhere as to essence and substance, is for Newton a nonbiblical God who is beyond nature (p. 121). To Newton's mind this view implies that God is present in his creation only as its ultimate and remote causal ground. But for Newton, God's power, substance, and essence are inseparable, and he acts where he is, namely, everywhere.

Newton also conceives God's immensity in a literal sense. In "De Gravitatione" he speaks of God's "quantity of existence" as infinite with respect to absolute space. This is clearly a conception of divine immensity.[13] By virtue of this notion Newton conceives the immensity of divine existence as unlimited, for it implies that if God should fail (*per impossibile*) to exist in any particular place, his quantity of existence would be diminished, and his individual presence thus limited. Newton attempts to preclude this consequence by denying the possibility "that a dwarf-god should fill only a tiny part of infinite space with this visible world created by him" (p. 121). In Newton's mind there is a close conceptual link between divine omnipresence and the conception of spiritual immensity. God is actually everywhere by virtue of his existence in infinite space, and in every place he wills everything that he thinks fit to choose. The immensity of God's omnipresence is manifested through his real presence in this created world and in the fact that he exists beyond it.

To distinguish God from space, Newton uses the Hebrew term *makom* in "Tempus et Locus" in speaking of divine omnipresence. The term is also used in the same context and for the same purpose in Newton's *Advertissement au lecteur*, which he sent to Des Maizeaux: "So when the Hebrews called God *MAKOM* place, the place in wch we live & move & have our being [they] did not mean that space is God in a literal sense."[14] Newton's use of *makom* is meant to convey the point that God dwells *in* space, not that space itself is a property of God's nature. This view is also conveyed in another passage. In the "classical scholia" intended for the unimplemented edition of the *Principia* (which was in preparation in the early 1690s—the same period as the present manuscript), Newton copies from Macrobius the sentence "The entire universe was rightly designated the Temple of God." He comments: "This one God they [the ancients] would have it dwelt in all bodies whatsoever as in his own temple, and hence they shaped ancient temples in the manner of the heavens."[15] It is clear that these figures are meant to invoke a sense of God's direct omnipresence throughout the created world. In Newton's use, *makom* is to some extent detached from its contexts in Jewish mystical and Cabalistic thinking. Moreover, his manner of using it differs in certain respects from that of Henry Moore, Samual Clarke, and Joseph Raphson.[16]

Newton's conception of God as an in-dwelling spirit omnipresent in the world is not pantheistic. That this is his view is clear in a manuscript intended for an unimplemented second edition of the *Principia* in the 1690s. Newton writes: "Those ancients who more rightly held unimpaired the mystical philosophy as Thales and the Stoics, taught that a certain infinite spirit pervades all space *into infinity*, and contains and vivifies the entire world. And this spirit was their supreme divinity, according to the Poet cited by the Apostle. In him we live and move and have our being."[17] Also a footnote in the General Scholium to the phrase "in him are all things contained and moved" refers to Cicero, Virgil, Philo, Aratus, and to "the sacred writers" from both the New and the Old Testaments.[18] The citations to their writings that Newton gives in each case reveal commitment to the notion of a divine mind, spirit, or life permeating and interpenetrating the cosmos, but in no way identified with space, time, or creation.

I have covered this material in detail to give as clear a picture as possible of Newton's theological conception of the nature and existence of God. "Tempus et Locus" also deals with the traditional attributes of divinity. God is said to be a necessary being, necessarily existing in all times and places. Moreover, Newton mentions three central attributes: eternity, omniscience, and omnipotence. It is not these characteristics that he is concerned to stress, however. God's nature is, of course, simple and indivisible. But more important in Newton's mind is the conception of God as an agent who is preeminently free to act in accordance with the perfections of his nature; he is truly alive, the maker and sustainer of life; he is an intelligent being who brings about "all things that are best and accord most with reason"; he has an immediate understanding and control of all things "just as the cognitive part of man perceives the form of things brought into the brain"; he is directly in contact with things that exist together with him in omnipresent space. It is these characteristics that make God a "most perfect being," and an agent best able to produce the great variety and design of creation (p. 123). In other words, it is God's actions in the real world that command our respect and demand true worship. This is the powerful Lord and God of the General Scholium, who exercises direct dominion over the constitution and governance of his creation. For a being "however perfect, without dominion, cannot be said to be Lord God."[19] Only a being that in fact has dominion over heaven and earth and all its creatures can signify the title of Lord or King, and thus command attitudes of obedience analogous to those of servants toward their earthly master, though in a much higher degree.

This conception of God's nature is further clarified by the Yahuda manuscript:

> To celebrate God for his eternity, immensity, omnisciency, and omnipotence is indeed very pious and the duty of every creature to do it according to his capacity, but yet this part of God's glory as it almost transcends the comprehension of man so it springs not from the freedom of God's will but the necessity of his nature—the wisest of beings required of us to be celebrated not so much for his essence as for his actions, the creating, preserving, and governing of all things according to his good will and pleasure.[20]

It is God's actions, then, emanating from his free and omnipresent will, that are manifest in his works, not the transcendent perfections of his essential nature. If the mind concentrates on the abstract features of God's infinite nature, it will lose its drift, according to Newton. It will fail to honor God for his direct and purposive dominion over the natural world. This is not to say, of course, that Newton rejects a metaphysical approach to God's nature. We must, however, guard lest such musings mask the significant aspect of God's nature—that he is a lordly master who is worthy of servants in virtue of the goodness and wisdom of his intelligent actions.

From all these considerations it is clear that Newton's primary motive is to make God comprehensible to the mind and worthy of attitudes of worship. Strictly conceived, God ought not "to be worshipped under the representation of any corporeal thing"; nevertheless "all our notions of God are taken from the ways of mankind by a certain similitude, which, though not perfect, has some likeness" to us. Accordingly, in the General Scholium, as in the manuscript under discussion, Newton insists that we know God best from the design of his creation, which reveals intrinsically his intentions and actions. By virtue of these characteristics "we reverence and adore him on account of his dominion" while "we admire him for his perfections."[21]

ii. Divine Existence and the Doctrines of Infinite Space and Time

The theological picture that emerges from Newton's religious sensibility is this: God is a being who always has existed and always will exist. His essential nature remains the same: he still is what he was and will continue to be what he is. On this view it is natural enough to think of God's existence as uncaused, such that there is never a time at which his being is preceded by anything else. For Newton this also means that God's nature is causally independent of any conditions, states of affairs, and circumstances external to his nature. Divine existence is explained solely by reference to God's previous existence, since existing omnitemporally is an essential fact pertaining to God's unending and permanent duration.

Furthermore, Newton does not hold that God is identical with Wisdom, Power, and Goodness, conceived as intrinsic attributes of his essential nature.[22] According to Newton's theory of perfection, characteristics such as being wise, powerful, and good are attributes that divine nature instantiates in the highest and most complete manner, namely, infinitely, where infinity is understood as a transcendental characterization not subsumable under the Aristotelian categories. Thus, to say that God is infinitely wise is to characterize the view that he cannot but act wisely. Nor can anything prevent God from acting in any way that is appropriate in this respect to his nature. For God can never change, nor be changed, with respect to the attribute of omniscience.[23]

On the assumption, therefore, that God possesses his perfections infinitely, nothing can prevent him from exercising his essential abilities. For all created things, including the uncreated natures of infinite space and time, are inferior to God's intrinsic perfections, since "by reason of its eternity and infinity space will neither be God nor wise nor powerful nor alive, but will merely be increased in duration and magnitude, whereas God by reason of the eternity and infinity of his space (that is, by reason of his eternal omnipresence) will be rendered the most perfect being."[24] The same line of reasoning applies to divine immutability. Unlike Spinoza, who is able uniquely to specify divine immutability *in alio*, or by virtue of God's intrinsic nature alone, Newton adopts a position that obliges him to consider the condition of God's existence in relation to the nature of other existents.[25] On this view, God is an immutable being in that there is nothing whatsoever that can *causally* change or in any way affect his defining nature and attributes. Thus, God is fully able to be identically one and the same person by virtue of his defining attributes at all possible and actual times; for there is nothing whatsoever by virtue of which he can be caused to be otherwise than he is. And Newton concludes in the light of his theory of perfection and infinity that "not everything eternal and infinite will be God, nor will God be prevented from the eternal and infinite exercise of his omnipotence in the creating and governing of things, by the imperfect nature of created things."[26]

For Newton infinity is a character that can be predicated of different sorts of things. It is a way of characterizing the mode of instantiation of first-level properties insofar as they allow something to be the highest exemplification of its kind. In itself infinity is neither a perfection nor an imperfection, nor is it defining of anything's nature. For "infinity is not a perfection except when it is attributed to perfections. Infinity of intellect, power, happiness, and so forth, is the height of perfection; but infinity of ignorance, impotence, wretchness, and so on, is the height of imperfection; and infinity of extension is so far perfect as that which is extended." Accordingly, to conceive spatial extension as infinite is not to conceive space as constituting "God because of the perfection of infinity."[27]

Newton's reasoning turns on the notion that things exemplify the nature that specifies the class or kind to which they belong. Furthermore, things can be conceived as either perfect or imperfect in acordance with the nature that they exemplify. In Newton's view there are levels of reality, and difference in level is defined in terms of the specific nature of the kind to which things belong, whether the reality in question is a perfection or an imperfection. A thing is infinite of the kind if it manifests in an exemplary manner the nature of that kind, but "no thing is by eternity and infinity made better or of a more perfect nature" ("Tempus et Locus," p. 120).

So, according to Newton's scheme of predication, a thing can instantiate one and the same nature infinitely (that is, completely) which another thing possesses deficiently. Thus, both God and we can be said to possess one and the same type of a given ability, except God possesses it infinitely, whereas we, in comparison, possess it deficiently. There are serious difficulties in this position. Standardly in theological thinking the notion that God has absolutely every perfection, such that he is uniquely one and the same being, is justified by arguing that the possession of every perfection implies supremacy and that supremacy implies uniqueness; otherwise there can be more than one supreme being. But Newton's account of the infinity of God's perfections (the notion that he possesses to the highest possible degree what other beings instantiate deficiently) provides no basis for ruling out the possibility that other beings might equally well exemplify the same perfections. Nor does his view that God transcends causal change help; for again it does not show necessarily that only *one* unique individual satisfies the claim. Here then is another example of Newton's anthropomorphic conception in which he holds that the same ontological principles apply to God that apply to finite things. And it raises the same difficulty that is present in his view that the composition of essence and existence that holds of God holds also of finite things.

The difficulty is this: How can Newton defend the unity and the uniqueness of divine being? I want to approach the larger framework of this issue by first considering if and how the dependence of space and time on God is causal. This question in turn is closely related to Newton's association of space and time with God's existence and also to the extent to which he "distances" existence from God's essence and substance.

Now, in my earlier study I argue that space, time, and existence function in Newton's divine ontology as "transcendental" predicates. In draft sheets intended for Des Maizeaux he tells us this about these special predicates:

> The Reader is desired to observe, that whenever in the following papers through unavoidable narrowness of language, infinite space or Immensity & endless duration or Eternity, are spoken of as *Qualities* or *Properties* of the substance such is Immense or Eternal, the terms *Quality* & *Property* are not taken in that sense wherein they are vulgarly, by the writers of *Logick* & *Metaphysicks* applied to matter; but in such a sense as only implies them to be modes of existence in all beings, & unbounded *modes* & consequences of the existence of a substance which is really necessarily & substantially Omnipresent & Eternal; which existence is neither substance nor a quality, but the existence of a substance with all its attributes, properties & qualities, & yet is so modified by place & duration that those modes cannot be rejected without rejecting the existence.[28]

Of this and cognate drafts I observed three things. First, that Newton is claiming that existence is not a real attribute or quality of the defining nature of substance. Second, that he connects space and time with the concept of actuality, that is, with the actuality of existence. And last, that Newton conceives these two doctrines as applying to divine existence itself.

Now, in order to secure this interpretation I argued that Newton was committed to the following basic perspectives:

(1) Newton tells us that "infinite space or Immensity" and "endless duration or Eternity" are not "*Qualities* or *Properties*" of God's defining nature, because they do not inhere in divine substance in the manner of the properties or accidents of matter. In other words, these phrases denote transcendental features because they fall under none of the ten Aristotelian categories nor under the fifth of Porphyry's predicables, accident. In another of the draft variants for Des Maizeaux he says that these phrases should be taken "in such a sense as if the Predicaments of *Ubi* & *Quando* should be called qualities or properties when applied to the existence of a being which is omni-

present & eternal." That is, when God is declared immense, or everywhere with respect to space, and eternal, or of unending duration in time, these are not properties, let alone specific properties, that are ascribed to divine nature. The phrases refer, rather, to the manner of God's existence. "*Ubi* & *Quando*" answer two general questions: Where does God exist? "Everywhere" in respect to space. When does God exist? "Always" in respect to time. *Ubi* & *Quando* are therefore "transcendental" predicates in the precise sense that they refer to conditions of existence that every actually existing thing, God included, must satisfy. This line of reasoning is present in the thought of many thinkers, including Tommaso Campanella, Francesco Patrizi, Pierre Gassendi, and Walter Charleton, each of whom argues that space and time are infinite and presupposed by the items in Aristotle's categories, and are thus the general conditions through which the actuality of any existing thing, God included, must be understood.[29]

(2) I was struck by the fact that Newton is at pains to stress that "existence is neither a substance nor a quality," but rather an irreducible feature of all individuals insofar as they are actual, where to refer to a thing's actuality is not at all like speaking of the properties of its defining nature. I was also struck by the fact that Newton goes on to say that anything's existence "is so modified by place & duration that these modes cannot be rejected without rejecting the existence." Thus, to exist is to exist in the general order of the nexus of space and time in a manner appropriate to the nature of the thing in question. So in speaking of God's existence, "the existence of a substance which is . . . substantially Omnipresent & Eternal," we refer to God's state of being actual with respect to the infinity of space and time. Thus, to be an actual being with respect to infinite space and time is an inseparable fact about divine existence. And when Newton speaks of space and time as "unbounded *modes* & consequences" of God's existence, he means that they are irreducibly associated with the necessary existence of God's eternal and omnipresent being.

Now, these perspectives on space, time, and existence seem to me to inform Newton's opening statement in "Tempus et Locus": "Time and Place are common affections of all things without which nothing whatsoever can exist. All things are in time as regards duration of existence, and in place as regards amplitude of presence. And what is never and nowhere is not in *rerum natura*" (p. 116). Here Newton makes a distinction between affections that characterize all the sorts of physical things and, by implication, those that are specific to various sorts and kinds. But there is another distinction that Newton has in mind. Common affections of course apply universally to all things, but time and place for Newton are special sorts of common affections. Notice that he stresses the phrases "duration of existence" and "amplitude of presence" in characterizing the association of time and place with the actuality of anything's existence. This implies that time and place specify anything's actual existence, in contrast to properties that inhere in a thing's specific nature. Thus, Newton probably has in mind a traditional distinction among affections: those that characterize things by virtue of their nature, and those that pertain universally to their sheer existence alone. Moreover, common affections of this latter sort are often categorized as external affections, again on the ground that they are not specific to any particular sort or kind. Thus, individual things not only endure in time; they are present in the same or different places through time. In "De Gravitatione" (c1668) the same line of reasoning is applied explicitly to God's "quantity of existence." Newton tells us that the "quantity of existence of each individual [being] is denominated as regards its amplitude of presence and its perseverance in existence. So the quantity of existence of God is eternal in relation to duration, and infinite in relation to the space in which he is present."[30] It is clear that Newton thinks of space and time as "common affections" of God's "quantity of existence." But in saying that space and time are "affections of being in so far as it is being," he is not saying that being in itself entails duration and extension in space (that is, that to be is to be extended and to endure). Again, he invokes a distinction between affections that characterize natures as natures and those that pertain to the fact that a natured individual exists. Of the latter kind are space and time.

Carriero is not happy with this account of why Newton associates space and time with God's existence rather than with his substance. Nor is he happy with reading Newton's claim in the General Scholium that God is not eternity and infinity but only eternal and infinite to mean that Newton is breaking with the traditional view of God as strictly identical with all his attributes, existence included. I am not happy with the implications of this reading either. It raises in acute form the problem of the unity of divine being.

I will come to this issue and the implications for my interpretation shortly. In order to approach the problem of unity, let me briefly consider Carriero's alternative interpretation. He concentrates on a passage from Newton's defender Clarke in which Clarke says that to claim God exists in space and in time means only that God is *"Omnipresent* and *Eternal,* that is, that *Boundless Space and Time* are necessary *Consequences* of his Existence." Carriero interprets the last phrase on space and time as being in apposition with the claim that God "is *Omnipotent* and *Eternal*." To Carriero this suggests two things: (1) that Clarke and Newton believe that an omnipresent existent is one whose existence necessarily produces boundless space, and an eternal existent is one whose existence necessarily produces boundless time; and (2) when Clarke and Newton claim that God is not identical with boundless space and time, they are not denying that he is identical with his attributes *omnipresent* and *eternal*. Their claim means, rather, that God is "not identical with the necessary causal consequences of those two attributes, boundless space and time." Thus, for Carriero infinite space and time are necessary emanations from a necessary being, and are therefore causally dependent on God.

There are two points, then, on which Carriero and I disagree. (1) Contrary to my view he believes Newton holds that God *is* omnipresence and eternity, the traditional view that God is *one* with his defining characteristics. (2) Given this conception, Carriero goes on to suggest that space and time are necessary *causal* productions of an eternal and omnipresent being. Accordingly, what God is *not* identical with on this view is the necessary *causal* consequences of these two attributes, namely, boundless space and endless time.

As to the first point of disagreement, I find it difficult to square Carriero's view with Newton's theory of divine predication. The evidence I adduce seems to indicate that Newton views God's omnipresence in relation to the "infinity of his space" and his existence in terms of the omnitemporality of his duration. On this view God's existence differs from finite existence in that it is infinite with respect to space and unending with respect to time. Thus God is not to be understood Platonically either as infinity itself or as eternity itself. The point is put succinctly in the General Scholium: As God "is not eternity and infinity, but eternal and infinite" so "he is not duration or space, but endures and is present."[31] Closely related to this is Newton's conception of space and time as items outside the categories; indeed, they are the invariant conditions necessary for the *actual* existence of anything, God included.

This brings me to the second point. When Newton says that space and time are consequences of God's existence, Carriero reads this claim as stating a *causal* relation between God's being and what that being produces necessarily. It is not clear, however, how Carriero understands the relation; for example, is it efficient or formal causation? It will be useful, then, to explore this issue briefly as it bears on the question of the ground of divine unity.

In my earlier study I agued that the relation between divine being and the infinity of space and time is one of ontic dependence. I now think it can be seen (in a curious sense) as a causal dependency, and, moreover, one that has a legacy in theological and philosophical thought. Noticing that Newton claims that space "is, as it were, an emanative effect of God," and also an affection of being qua being, I suggest that a possible influence is Henry More, who views space as an "effectus emanativus" of divine being and also an attribute of "ens quatanus ens." More defines "an emanative effect" as "coexistent with the very substance of that which is said to be the cause thereof." And "an emanative cause" is understood as "such a cause as merely by Being, no other activity or causality interposed, produces an effect."[32] Noting also that More denies that any action or active causal efficacy obtains between an emanative cause and its effect, I argued that the relationship for More is ontic rather than causal, and that Newton also views the matter in this way.

Interestingly enough, there is medieval background for reading this relationship under the rubric of *efficient* causation. In the writings of Duns Scotus and Robert Grosseteste we find the idea that divine power is causally prior to its acts.[33] Moreover, this type of causal priority holds between things existing at the same time, or, in the case of God and his acts, together in eternity. This view of course involves the commonplace medieval conception that causes need not precede their effects. Indeed, not only need they not necessarily precede their effects, but they need not exist in time at all. The cause, however, is naturally prior to its effects both temporally and eternally since the effect has no being without its cause. But causal priority among eternal things does not require the *creation* or *production* of the effect by the cause. The effect has being just because the cause *is simpliciter*, but the converse does not hold. Augustine's foot eternally embedded in dust, and thus eternally causing its footprint, is an example of this relationship.

There is medieval precedent, then, for speaking of eternal *and* efficient causes. And something of this sensibility may be reflected in More's notion of an "emanative cause," and similarly in Newton's view of the relationship between divine existence and the reality of space and time. But since the notion of an eternal and efficient cause does not involve any activity, production, creation, or active efficacy between it and its effect, it is difficult to distinguish natural or ontic dependence in these contexts from the notion of causal dependence between eternal things.

Now, both Carriero and I want to argue that Newton associates space and time with God's existence in an attempt to distance them from divine essence. For Carriero, Newton makes this maneuver because he thinks the association of space and time with divine existence best fits his conception of space and time as necessary emanations from a necessary being. For me it is because infinite space and time characterize best Newton's view of the nature and manner of divine existence as such, as well as his belief that God is continuously present to creation in a dynamic and providential way. Carriero's interpretation has the difficulty that there is no compelling reason why a necessary being (one who *is* both eternity itself and infinite presence itself) must necessarily generate space and time from the necessity of its being. This is an emanationist model of creation which holds that a necessary being necessarily creates by emanation from its being. But why should Newton's Christian and voluntarist God necessarily create anything, let alone externalize space and time as necessary emanations from its essential attributes?

But my position also harbors a difficulty. To claim that Newton believes space and time to be essential conditions for the *actuality* of all existing things, God included, has the consequence that the unity of God's essence and existence is threatened. Newton can hold that the existence of a finite thing is grounded in its nature because God has actualized the concrete existence of that nature. It is thus unified by God's creation act. But what unifies God's existence and essence? The claim that God exists eternally, in the sense that there is no time at which his existence can fail, does not ground the necessity of his existence. The modal notion that God can at no time fail to exist cannot itself be grounded in the claim that God exists omnitemporally. From the temporal "exists at all the times there are" one cannot derive the modal "cannot not exist at any time." That God cannot not exist is a claim about his nature as such, not a claim about sheer omnitemporal existence. Thus the conception that God's existence is eternal and omnitemporal still allows that a contingent relation obtains between divine essence and existence. It seems, then, that to preserve divine unity, Newton must fall back on the traditional view that God's existence is a necessary perfection of his nature if God is to be conceived as a necessary being. But then why should that necessary nature need to externalize space and time as necessary consequences of its essential attributes, and why should its actuality demand existence with respect to the infinity of space and time?

Notes

1. J. E. McGuire, "Existence, Actuality, and Necessity: Newton on Space and Time," *Annals of Science* 35 (1978), 463–508.

2. Frank E. Manuel, *The Religion of Isaac Newton* (Oxford: Clarendon Press, 1974), chap. 2.

3. J. E. McGuire, "Newton on Place, Time, and God: An Unpublished Source," *British Journal for the History of Science* 11 (1978), 114–129. Page references to this article will be given in the text.

4. Alexandre Koyré and I. Bernard Cohen, "Newton and the Leibniz-Clarke Correspondence," *Archives internationales d'histoire des sciences*, nos. 58–59 (1962), 101.

5. Issac Newton, *Mathematical Principles of Natural Philosophy*, ed. Florian Cajori (Berkeley: University of California Press, 1960), p. 545. *Isaac Newton's Philosophiae Naturalis Principia Mathematica: The Third Edition (1726)*, assembled and edited by Alexandre Koyré and I. Bernard Cohen, 2 vols. (Cambridge, Mass.: Harvard University Press, 1972), II, 759.

6. A. Rupert Hall and Marie Boas Hall, *Unpublished Papers of Isaac Newton* (Cambridge: Cambridge University Press, 1962), p. 357. Koyré and Cohen, op. cit., p. 97.

7. See J. E. McGuire and Steven K. Strange, "An Annotated Translation of Plotinus, Ennead 111.7 on Eternity and Time," *Ancient Philosophy* 8 (1989), 251–271.

8. H. McLachlan, *Sir Isaac Newton: The Theological Manuscripts* (Liverpool: The University Press, 1950), p. 56.

9. Op. cit. (n. 2), p. 101.

10. University Library, Cambridge, Add. 3965, sec. 13, folio 542ʳ.

11. I. Bernard Cohen, "Isaac Newton's *Principia*, the Scriptures, and the Divine Providence," in *Philosophy, Science, and Method*, ed. Sidney Morgenbesser, Patrick Suppes, and Morton White (New York: St. Martin's, 1969), p. 528.

12. Op. cit. (n. 5), p. 545.

13. Op. cit. (n. 6), p. 104.

14. Op. cit. (n. 4), p. 101.

15. Gregory MS, 247, Library of the Royal Society. See J. E. McGuire and P. M. Rattansi, "Newton and the 'Pipes of Pan,' " *Notes and Records of the Royal Society of London* 21 (1966), 108–143.

16. See A. Koyré, *From the Closed World to the Infinite Universe* (New York: Harper & Row, 1958), chaps. 6 and 8; and op. cit. (n. 4), for the view that Cabala has little real influence on Newton. For a careful and comprehensive discussion of *makom* and *simsum* in the writings of More and Raphson and the relation of these writers to Newton, see Brian P. Copenhaver, "Jewish Theologies of Space in the Scientific Revolution: Henry More, Joseph Raphson, Isaac Newton, and Their Predecessors," *Annals of Science* 37 (1980), 489–548.

17. Portmouth Collection MS, Add. 3965, 12, f.269 VLC. See also McGuire and Rattansi, op. cit. (n. 15), p. 120.

18. Op. cit. (n. 5), pp. 545, 759.

19. Ibid., pp. 544, 759.

20. Op. cit. (n. 2), pp. 21–22.

21. Op. cit. (n. 5), pp. 545–546, 759–760.

22. Ibid., p. 545.

23. See McGuire, op. cit. (n. 1), for a discussion of this conception.

24. Op. cit. (n. 3), p. 118.

25. See op. cit. (n. 1), for a discussion.

26. Op. cit. (n. 3), p. 120. See also "De Gravitatione" in Hall and Hall, op. cit.

27. See "De Gravitatione," pp. 102–103.

28. Op. cit. (n. 4), p. 101.

29. See McGuire, op. cit. (n. 1), for discussion of these thinkers.

30. Op. cit. (n. 6), p. 103.

31. Op. cit. (n. 5), pp. 545, 759.

32. See McGuire, op. cit. (n. 1), for a discussion of More's views.

33. John Duns Scotus, *God and Creatures: The Quadlibetal Questions,* trans. Felix Alluntis and Allan B. Wolter (Princeton: Princeton University Press, 1975), question 19, art. 11, 19.27, 19.28, 19.29. John Duns Scotus, *De Primo Principia,* trans. Allan B. Wolter (Chicago: Franciscan Herald Press, 1966), 2.29, 2.31, 2.32, 2.33. Robert Grosseteste, *Commentarius in VIII Libros Physicorum Aristotelis,* ed. Richard C. Dales (Boulder: University of Colorado Press, 1963), Liber Quartus, 96–97.

Chapter 7

Newton on Space and Time: Comments on J. E. McGuire

John Carriero

Space and time, for Newton, make up the absolute frame not only in which all physical transactions take place but through which all beings, God and spirits not excepted, are interrelated. In Aristotelian metaphysics, which was still quite influential in Newton's day, being is divided into substance and accident. It is not surprising that Newton should write in the course of a discussion concerning the nature of space: "Perhaps now it may be expected that I should define extension as substance or accident or else nothing at all."[1] Newton is adamant about the reality of space—that is, about space's not being "nothing at all"—and sees himself, in this regard, as opposing what "is usually believed":

> But it is usually believed that these spaces are nothing; yet indeed they are true spaces. Although space may be empty of body, nevertheless it is not in itself a void; and *something* is there, because spaces are there, although nothing more than that. ("De Gravitatione," p. 138)

> And much less may it [extension] be said to be nothing, since it is more thing-like than an accident, and more closely approaches the nature of substance. ("De Gravitatione," p. 99, McGuire's translation)

Does it follow, then, that space is either a substance or an accident? "By no means," writes Newton, "for it [extension] has its own manner of existence which fits neither substances nor accidents." Space is not an accident because it does not inhere in anything; space is not a substance because it lacks the activity and passivity that are assumed, at least implicitly, to accompany substance.[2]

Newton's account of space and time does not end here, however. He also writes that space and time are "affections of being *qua* being,"[3] that is, properties that belong to being as such. J. E. McGuire in "Predicates of Pure Existence: Newton on God's Space and Time,"[4] and in earlier work from which the present paper draws, especially

"Existence, Actuality and Necessity: Newton on Space and Time,"[5] offers an interesting (and, to my knowledge, the only systematic) treatment of Newton's thesis that space and time are affections of being qua being. Briefly, according to McGuire, Newton holds that space and time are general conditions of being which attach specially to a thing's existence. Newton, as McGuire understands him, couples this thesis with a further claim that what belongs to something's existence is really distinct from what belongs to its essence, which enables him to maintain that space and time, as affections of divine existence, are not attributes or qualities of the divine essence.

In my comments I take up two points. In the first section I argue that McGuire's interpretation on occasion shortchanges the reality that Newton attributes to space and time. In the second section I begin by offering an account of the relationship between Newton's claim that space and time are common affections and his view that space and time are real beings. Further, I argue that the distinction that McGuire sees Newton as drawing between a thing and its common affections is neither philosophically nor theologically coherent. Finally, I offer an interpretation of Newton's association of space and time with existence that does not involve him in this incoherence.

i. Space and Time as Real Beings

Viewing space and time as general conditions of being tends to pull one away from thinking of space and time as real beings. When they are so viewed, it is difficult to see why they should have any existence apart from the beings whose existence they condition. Consider a standard example of a property coextensive with being: unity. It does not exist on its own, apart from the oneness found in the individual things existing in the universe. On occasion, McGuire's preoccupation with Newton's claim that space and time are general conditions of being leads him to shortchange the reality of space and time. A case in point is his reading of Newton's remark that space is more like a substance than an accident:

> Nevertheless, he [Newton] does say that space (and by implica-
> tion time) "is more thing-like than an accident, and more closely
> approaches the nature of substance." This indicates that Newton
> has in mind a sense in which space is a substance. While it is not
> a substance in the sense of existing in itself, or in the sense of
> being the source of change in other things, it is nevertheless a
> distinct individual of which many things can be said. After all, in
> essence, space is fully actual and infinite extension to which
> many definite properties are truly ascribable. (EAN, p. 473)

McGuire apparently understands Newton's remark to rest on the fact that space can function as a logical subject of discourse.[6] But the context of the remark—it occurs in a discussion of whether space is a substance or an accident or "nothing at all"—makes it obvious that Newton's point about the near-substantiality of space is not logical but ontological. Even blindness would count as a substance in the sense outlined by McGuire: Homer's blindness, for example, "is a distinct individual of which many things can be said" and "to which many definite properties are truly ascribable." But surely, in this context, blindness should count as an *absence* of being, that is, as a "nothing." Further, McGuire's suggestion that space "is not a substance in the sense of existing in itself" is in conflict with an immediate consequence of Newton's argument that space is not an accident because it does not inhere in anything: if space is a real existent, and if it does not inhere in anything, then it must exist in itself.[7]

A second place where McGuire shortchanges the reality of space and time concerns their causation. If space and time are really existing beings, then they require some efficient cause that is responsible for their existing.[8] And, indeed, Newton describes them as "emanative effects" of God. But if we think of space and time as general conditions of being, then it is harder to think of them as having efficient causes. Again, a comparison with unity is helpful: one would not think of unity as being efficiently caused by the various "one" beings that exist. The general orientation of McGuire's reading leads him to understand Newton's statement that space and time are emanative effects of God in a way that deemphasizes the efficiency or activity that is involved in this relation. In his earlier writing he denied outright that efficient causation is involved in the emanation of space from God. Influenced by his view that "space is . . . conceived [by Newton] as the general condition required for the existence of any individual substance including its characteristics" (EAN, p. 481), McGuire wrote: "The relation between the existence of being and that of space is not causal, but one of ontic dependence. Newton is defining one condition which must be satisfied so that any being can be said to exist. In short, the phrase 'when any being is posited, space is posited' denotes an ontic relation between the existence of any kind of being and the condition of its existence" (EAN, p. 480).[9] More recently, McGuire has modified this position and is now willing to allow that emanation might count as an instance of efficient causation, but only in some protracted sense thereof. Relying on More's account of emanation in particular, he argues that "the notion of an eternal and efficient cause does not involve an activity, production, creation, or active efficacy, between it and its effect," which makes it "difficult to distinguish natural or ontic dependence in these contexts from the notion of causal dependence between eternal things" (PPE, section ii).

In the medieval tradition emanation is a kind of necessary, efficient causation. Such causation is necessary in the sense that if an emanative cause exists, it is impossible for its effect not to exist. Emanation is found in certain theories, opposed to creation, of the production of the universe. For example, Al-Farabi and Avicenna held that there is a First Necessary Being (God), from which emanates a first intelligence, from which emanates a second intelligence, from which emanates the form of the outermost celestial sphere.[10] Since it is impossible for an emanative cause to exist and its effect not to exist, an emanative cause is temporally coterminous with its effect and, in particular, not temporally prior to its effect.[11] Although emanation is primarily found in accounts of the production of the universe, the notion is also used in other situations where it is impossible for the cause to exist and its effect not to exist. The relation of a proper accident to its subject, as understood by Aquinas, is a case in point. A proper accident is something that, although not falling within the definition of its subject, is such that its subject never exists without it. *Risible* is a proper accident of a human being: although *risible* is not found within the definition of the essence of a human being (*rational animal*), it is a necessary concomitant of human nature. If risible does not belong to the essence of human being, what accounts for the fact that human nature is never found without risibility? According to Aquinas, the subject is "active cause" of its proper accidents, which causation he describes as "a flowing from" or "an emanation from" the subject's essential principles. Moreover, as a subject is never without its proper accident, Aquinas holds that this causation is instantaneous, comparing it to the dependence of color on light.[12]

This traditional understanding of emanation creates a presumption that when Newton writes that space and time are emanative effects of God, he is claiming that God efficiently causes space and time. According to McGuire, however, there is contrary textual evidence in both Newton and More. To begin with Newton:

> The implications of Newton's position can be seen in the following passage: "Space is of eternal duration and immutable nature, because it is an emanative effect of an eternal and immutable being. If ever space had not been, God at that time would have been nowhere, and hence he either created space later, where he himself was not present or else, which is not less ab-

surd to reason, he created his own ubiquity." Here again we see that space (and by implication time) are coeternal with the existence of God. On this assumption, infinite space and time cannot be caused by God's existence. For if anything is truly infinite it is uncreated. Moreover, the conception of efficient causation, where a cause is said to be prior to its effects, is incompatible with Newton's claim that God, space and time are coeternal. For it involves the absurdity (as Newton himself points out) of saying that God either existed nowhere before he efficiently created space and time, or that he continuously creates his own ubiquity. The latter consequence must also be rejected, since it conceives God as actively creating his own presence in infinite space and time. This, of course, is inconsistent with Newton's conception that infinite space and time exist coeternally with God. In any event, what would it mean to say that God eternally actualizes them? Newton's use of the phrase "emanative effect" is therefore misleading in so far as it connotes adherence to any form of emanationist causality.   (EAN, pp. 481–482)

McGuire's handling of Newton's argument is inexact. To begin with, consider the connection between the infinity of space and its being uncreated. It is true that many thinkers took infinity to be an attribute that belonged exclusively to God, and so for them "if anything is truly infinite it is uncreated." But Newton has his own understanding of infinity, developed precisely in order to allow that other beings besides God might be infinite:

> But I see what Descartes feared, namely that if he should consider space infinite, it would perhaps become God because of the perfection of infinity. But by no means, for infinity is not perfection except when it is an attribute of perfect things. Infinity of intellect, power, happiness and so forth is the height of perfection; but infinity of ignorance, impotence, wretchedness and so on is the height of imperfection; and infinity of extension is so far perfect as that which is extended.   ("De Gravitatione," pp. 135–136)

Moreover, Newton does not ground space's being uncreated in its being infinite in the passage that McGuire cites (nor does Newton do so elsewhere, to my knowledge); indeed, it is not even clear that Newton rules out the possibility that there might be found in the universe infinitely many solar systems like our own, the collection of which would surely be infinite and created.[13] A further problem is McGuire's movement from space and time's being *uncreated*, for which Newton does argue, to space and time's being *uncaused*, for which Newton does not argue. As we saw when we surveyed the doctrine of emanation, emanation is a kind of efficient causation which, in both its necessity and its temporal coincidence of cause and effect, is opposed to creation. Moreover, it is these very features of emanation that Newton emphasizes in the passage. Emanation, unlike creation, can underwrite space's eternity and immutability: "Space is eternal in duration and immutable, and this because it is the emanent effect of an eternal and immutable being." And emanation, unlike creation, can secure the necessary existence of space; thus, the passage continues: "Next, although we can possibly imagine that there is nothing in space, yet we cannot think that space does not exist, just as we cannot think there is no duration, even though it would be possible to suppose that nothing whatever endures" ("De Gravitatione," pp. 137–138).

Let's turn to More's views on emanation. Having correctly urged the likely influence of More on Newton, McGuire continues:

> More defines "an emanative effect" as "coexistent with the very substance of that which is said to be the cause thereof." And "an emanative cause" is understood as "such a cause as merely by Being, no other activity or causality interposed, produces an effect." Noting also that More denies that any action or active causal efficacy obtains between an emanative cause and its effect, I argued that the relationship for More is ontic rather than causal, and that Newton also views the matter in this way. (PPE, section ii)

This treatment of More's views, however, is imprecise. To begin with, McGuire simply overlooks the word *other* in More's account of an emanative cause; but obviously there is a great deal of difference between denying that any activity is involved in the relation of an emanative cause to its effect and denying that no activity is involved beyond that found in the cause's being. Since the key idea in emanation is that it is impossible for the cause to exist and the emanative effect not to exist, it is not surprising that More should suggest that in order to produce its effect, an emanative cause need do nothing besides whatever activity is involved in the cause's being or existing; it would be quite surprising if he were to write that there is no activity at all in the production of an emanative effect. Moreover, the surrounding texts place it beyond doubt that More has a relationship of efficient causation in mind. For example, he writes that an emanative effect cannot exceed "the virtues and powers of the Cause" and offers the following account of why an emanative effect necessarily coexists with its cause:

Axiome XVII

An Emanative Effect is coexistent with the very Substance of that which is said to be the Cause thereof.

This must needs to be true, because that very Substance which is said to be the Cause, is the adequate and immediate Cause, and wants nothing to be adjoyned to its bare essence for the production of the Effect; and therefore by the same reason the Effect is at any time, it must be at all times, or so long as that Substance does exist.[14]

There is no reason not to take More's and Newton's commitment to emanation at face value as embodying a relation of full-fledged efficient causation.[15]

The emanation of space and time from God holds certain consequences for their relation to God. On the one hand, the emanation of B from A precludes B's belonging to A's essence. This is because A cannot be an efficient cause of itself: Socrates cannot, for example, efficiently cause his being rational. On the other hand, B may still exist in A as an accident in subject. Thus, although *risible* emanates from human nature, it is also an accident of a human being. Thus, the emanation of space and time from God precludes their belonging to the divine essence; it does not, however, preclude their existing in God as accidents.

ii. Space and Time as Common Affections of Divine Being

Space and time are, for Newton, real beings, efficiently caused by God; their reality will not dissolve into that possessed by a mere logical subject. But Newton also views space and time as affections of being qua being. Does Newton's wanting to view space and time in these two ways mark a deep tension within his thought?

What is a common affection? To claim that unity is an affection of being as such is to claim that each thing, as a being, is one. Affections, moreover, are numerically multiplied by their possessors: it is an affection of Socrates, for example, as a being, to be one, and it is an affection of Plato, as a being, to be one; but Socrates' oneness, inasmuch as it is an affection belonging to him, is not the same as Plato's oneness, inasmuch as that belongs to Plato. Now, how should we understand Newton's claim that space and time are common affections? To say that space, for example, is an affection of Socrates would seem to be a way of saying that Socrates, as a being, must be in space. Thus, Newton writes: "Time and Place are common affections of all things without which nothing whatsoever can exist. All things are in time as regards duration of existence, and in place as regards amplitude of presence. And what is never and nowhere is not in *rerum natura*" ("Tempus et Locus," p. 117). Since the affection involves being in space, it seems to me to be more apt to refer to it as spatial locatability[16] instead of simply as space, as Newton does. I take him to be claiming, then, that spatial locatability is an affection of being as such, so that it is an affection of Socrates, as a being, to be locatable in space. Moreover, as with unity, Socrates' spatial locatability, inasmuch as it is an affection of Socrates, is not the same as Plato's spatial locatability, inasmuch as it is an affection of Plato. (Similar remarks are in order as regards time and temporal locatability.)

How does Newton understand the relationship between the affections of space and time—that is, spatial and temporal locatability—and the real existents space and time? Since there is no real existent in the case of unity, there is no immediate parallel with unity to guide us here. We could imagine, however, a "Platonic" view according to which there exists a one itself and other beings possess their oneness by virtue of their participation in the one itself. In this case there would be a structural similarity between the relationship of the affection one (that is, unity) to the real being one, on the one hand, and the relationship, as Newton conceives it, of the affections space and time (that is, spatial and temporal locatability) to the real beings space and time, on the other. For, although Newton does not understand a thing's spatial locatability in terms of its participation in space, he does understand such locatability in terms of its bearing a certain relation to space: "Space is an affection of being qua being. No being exists or can exist which is not related to space in some way. God is everywhere, created minds are somewhere, and body is in the space that it occupies; and whatever is neither everywhere nor anywhere does not exist" ("De Gravitatione," p. 136). In other words, for A to be spatially locatable is for A to bear one of the three relations—being everywhere, being somewhere, occupying—to space. Newton's conception of space as an affection and his conception of space as a real being are in a certain sense complementary: something acquires the affection space by virtue of being related in one of three ways to the real being space. When Newton's views on space and time are appropriately fleshed out, then, there is no tension between his claim that space and time are common affections and his claim that space and time are real beings. But Newton's use of the same word, *space*, for both the affection and the real being does give rise to a certain ambiguity. The question, for example, of God's relation to space becomes, in effect, two questions: First, there is a question concerning his relation to a certain real existent, effected by him, namely space; and second, there is a question concerning his relation to certain of his affections, namely his spatial locatability. (As before, similar remarks are in order as regards time and temporal locatability.)

The relation of space and time to God is a topic that is of particular interest to McGuire. (Since McGuire does not, at least explicitly, distinguish between space and time taken as common affections and space and time taken as real existents,[17] when summarizing his views I shall use the expressions *space* and *time* for both the affections and the real beings.) According to McGuire, space and time, for Newton, are not merely common affections but a particular kind of common affection which attaches to a thing's existence. What is the significance of this association with existence? McGuire sees the following passage from drafts of a paragraph that Newton wrote for the Des Maizeaux edition of the Leibniz-Clarke correspondence as holding the key:

> The Reader is desired to observe, that wherever in the following papers through unavoidable narrowness of language, infinite space or Immensity & endless duration or Eternity, are spoken of as *Qualities* or *Properties* of the substance wch is Immense or Eternal, the terms *Quality & Property* are not taken in that sense wherein they are vulgarly, by the writers of *Logic & Metaphysics* applied to *matter*; but in such a sense as only implies them to be modes of existence in all beings, & unbounded *modes* & consequences of the existence of a substance which is really necessarily & substantially Omnipresent & Eternal; Which existence is neither a substance nor a quality, but the existence of a substance with all its attributes properties & qualities, & yet is so modified by place & duration that those modes cannot be rejected without rejecting the existence.[18]

According to McGuire, Newton is claiming here that a thing's existence is really distinct from its essence (and its accidents as well).[19] This distinction opens up new ontological possibilities for the relation of space and time to God's essence; it provides Newton with the means to hold that space and time neither belong to the divine essence nor are accidents of his nature but are instead modes or affections of his really distinct existence. As modes of existence, one would expect, space and time should inherit existence's distinctness from essence, an expectation that McGuire takes to be borne out by

Newton's claims in the General Scholium that space and time are not God: "He is not eternity and infinity, but eternal and infinite; he is not duration and space; but he endures and is present."[20] Further, as infinity and eternity are attributes of God, this last claim, as McGuire reads it, embodies a departure from traditional theology: Newton "opposes those theologians who hold that God's attributes are strictly one and the same with Divine essence and existence, as well as identical among themselves" (EAN, p. 479).

It must be admitted, I think, that the sense of Newton's remarks in the Des Maizeaux passage, particularly of those that touch on existence and its relation to substance, does not leap from the page. In particular, when Newton writes, "Existence is neither a substance nor a quality, but the existence of a substance with all its attributes properties & qualities," it is not obvious that he is making the ontological point that a substance and its accidents are really distinct from its existence. I wish to argue, first, that there is no reason for Newton to make such a distinction; second, that it is philosophically and theologically incoherent to draw a real distinction in this way between a thing's essence and its existence; and finally, that it is possible to arrive at a satisfying reading of the Des Maizeaux drafts without reifying the distinction between essence and existence found there.

Part of what is driving McGuire's interpretation is a desire to square Newton's claim that space and time are, as affections of being qua being, affections of God, with the statement in the General Scholium that God "is not duration and space."[21] We find a similar denial in Clarke's third response to Leibniz: "Infinite space, is not God."[22] If, however, we keep in mind the distinction, discussed at the beginning of this section, between space and time taken as God's affections and space and time taken as real beings, it is not so obvious that these texts constitute denials that God is identical with certain of his affections. Newton and Clarke may simply be denying that God is identical with the real beings space and time and not that he is identical with his affections of spatial and temporal locatability. Indeed, this seems to me the most natural reading of their remarks. Consider the passage from the General Scholium in its surrounding context:

> He is eternal and infinite, omnipotent and omniscient; that is, his duration reaches from eternity to eternity; his presence from infinity to infinity; he governs all things, and knows all things that are or can be done. He is not eternity and infinity, but eternal and infinite; he is not duration and space; but he endures and is present. He endures forever, and is everywhere present;

and, by existing always and everywhere, he constitutes [*constituit*] duration and space.   (Newton, II, 761, p. 478)[23]

Newton glosses God's being eternal here as "his duration [reaching] from eternity to eternity" and his being as "his presence [reaching] from infinity to infinity." Clarke offers a similar, though more causally oriented gloss, in one of his subsequent letters: "When, according to the analogy of vulgar speech, we say that he [God] exists in all space and in all time; the words mean only that he is omnipresent and eternal, that is, that boundless space and time are necessary consequences of his existence" (C.V.36–48). God's being "omnipresent and eternal" is used by Clarke in apposition with his having "boundless space and time" as "necessary consequences of his existence." Implicit in these glosses is, in the case of Newton, a distinction between God's *being eternal* (that is, having a duration reaching from eternity to eternity) and God's eternity (the time through which his duration reaches), and God's *being infinite* (his having a presence reaching from infinity to infinity) and *his infinity* (that is, the infinite expanse through which his presence reaches); and, in the case of Clarke, between God's being "omnipresent and eternal" (that is, his having "boundless space and time" as "necessary consequences of his existence") and the boundless space and time themselves that are consequences of his existence. But, clearly, it is having a "duration [which reaches] from eternity to eternity" and a "presence [which reaches] from infinity to infinity (or, less obviously, having an existence of which "boundless space and time are necessary consequences")[24] that constitutes God's spatial and temporal locatability, that is, space and time taken as affections of God. Further, these passages do not suggest that God is not identical with his having a "duration [which reaches] from eternity to eternity," and so on, and so do not suggest here that God is not identical with certain of his affections. These texts do not give us any reason, then, for believing that Newton either denies that God is identical with certain of his affections or affirms that God has certain affections that lie outside the divine essence.

Not only is there no reason to reify Newton's distinction between essence and existence, but doing so involves intractable philosophical and theological difficulties. To begin with, the fact that A is a common affection of B does not of itself invite the thought that A is really distinct from B. One would be hard-pressed to find some real distinction between Socrates (or his essence) and his unity, and then to begin attributing properties to his unity as opposed to his substance. Now, according to McGuire, what is supposed to ground a real distinction here is that existence and being in space and time are not simply common affections but a special kind of "external" common affection. But what is an external affection? McGuire's explication of this notion is based on the following passage from "Tempus et Locus": "Time and Place are common affections of all things without which nothing whatsoever can exist. All things are in time as regards duration of existence, and in place as regards amplitude or presence. And what is never and nowhere is not in *rerum natura*." McGuire writes concerning this passage:

> Newton probably has in mind a traditional distinction among affections: those that characterize things by virtue of their nature, and those that pertain universally to their sheer existence alone. Moreover, common affections of this latter sort are often categorized as external affections, again on the ground that they are not specific to any particular sort or kind. Thus individual things not only endure in time, they are present in the same or different places through time.   (PPE, section ii)[25]

Simply observing that there is a "traditional distinction among affections," according to which some affections pertain to a thing's essence and some to its "sheer existence," fails to address the crucial issue here as to whether these "external" affections are considered to be ontologically external—that is, whether such affections were understood to be really distinct from the substance and accidents of the thing to which the affections belong. More troubling still is that his exposition of the distinction makes it hard to see why they would be regarded as ontologically external. He tells us that the reason for categorizing certain affections as external is that "they are not specific to any particular sort or kind." But why wouldn't the generality here envisioned hold good for unity? (Indeed, why wouldn't it hold good for all common affections—that is, affections of being qua being?) To be sure, unity, because of its generality, is not included in the definition of the specific nature of a thing: the essence of a human being is defined as "rational animal" and not as "one rational animal." Surely, however, the explanation of this is not that some real distinction obtains between a human being and his or her oneness but rather that, since unity belongs to all beings as such, it is not worth mentioning explicitly—or better still, that unity is already implicitly included in the definition of a human being, inasmuch as the definition, when fully analyzed, would read "rational, sensitive, living, corporal substance," and substance, as one of the two kinds of real being (the other is accident), involves the possession of all the affections that belong to being as such. Finally, it is unclear what bearing the observation that McGuire offers at the end of the passage—namely, that "things not only endure in time, they are present in the same or different places through time"—is supposed to have on the question of the externality of certain common affections. To be sure, the fact that A is found in other times besides time t_o and in other positions besides x_o, y_o, z_o could be taken to show that it does not belong to A's essence to be at t_o or at x_o, y_o, z_o: but no one is claiming, I take it, that it belongs to being as such to be at time t_o or in position x_o, y_o, z_o, but rather that it belongs to being as such to be in time and in space.

It might be felt that there is an important dissimilarity between unity, which I have been using as a paradigm of a common affection, and existence, a dissimilarity that might serve as the basis for the real distinction that McGuire supposes Newton to draw between essence and existence. Whereas it is necessary that Socrates, as a being, be one being, it is only contingent that he exist. And there were thinkers (most notably, in the Latin West, Aquinas) who held that there is, in the case of finite or contingently existing things, a distinction, grounded in reality, between essence and existence. Although the precise nature of the distinction is controversial, it is clear that the basic idea behind the distinction is to mark a finite thing's need to receive its existence from God. But, on McGuire's interpretation, Newton distinguishes between essence and existence not only in the case of finite things but also in the case of God himself; and as Newton regards God's existence as necessary, it is evident that a distinction grounded on the contingency of existence cannot serve his purposes.

Finally, the position that McGuire attributes to Newton is theologically untenable. According to McGuire, Newton draws two related distinctions with God's being: first, he holds that God's essence and existence are really distinct; second, he holds that God and his attributes are really distinct. Both claims are incompatible with the simplicity traditionally attributed to God: according to traditional theology, although one could draw a distinction of reason between God and his existence or attributes (that is, what we might think of today, somewhat anachronistically, as a conceptual distinction), there is no basis in reality for any such distinction. It would mark a particularly radical departure from the tradition to hold that there is some ontologically based distinction between God's existence and his essence, for, as Harry Wolfson observes, the thesis that God is identical with his existence serves as "the main principle from which arise all the negations and affirmations about the divine nature."[26] Moreover, the difficulty with a composite God is obvious: if God is composite, there must be some principle prior to God that is responsible for his composition; but this is incompatible with God's primacy.

The position that McGuire attributes to Newton is not, I conclude, sufficiently coherent for us to believe that Newton held it. Is there some other way to read the Des Maizeaux passage without having him drive an ontological wedge between the divine existence and the spatial and temporal locatability associated with it, on the one hand, and the divine essence, on the other? It is clear that the Des Maizeaux paragraph is designed to combat some misimpression that Clarke's use of "property" in his correspondence with Leibniz might leave. So then, how might Clarke's characterization of space and time as properties of God, if taken at face value, present a distorted picture of the Newtonian position?

Whatever else Newton is worried about, he is not worried that Clarke may have given the impression that space and time (the real beings) are part of the divine essence. I noted at the end of the first section that Newton's view that God is the emanative cause of space and time precludes space and time from being found in the divine essence, and Clarke makes it sufficiently clear that space and time are not God because they are causal consequences of God:

> Space is not a being, an eternal and infinite being, but a property, or a consequence of the existence of a being infinite and eternal. Infinite space, is immensity: but immensity is not God: And therefore infinite space, is not God. (C.III.3)

> Space is not a substance, but a property; and if it be a property of that which is necessary, it will consequently (as all other properties of that which is necessary must do,) exist more necessarily, (though it be not itself a substance,) than those substances themselves which are not necessary. Space is immense, and immutable, and eternal; and so also is duration. Yet it does not at all hence follow, that any thing is eternal *hors de Dieu*. For space and duration are not *hors de Dieu*, but are caused by, and are immediate and necessary consequences of his existence. And without them, his eternity and ubiquity (or omnipresence) would be taken away. (C.IV.10)[27]

Rather, the problem is that Clarke's "uncorrected" text leaves the impression that space and time are straightforward *propria* of God— that is, necessary accidents of the divine being.[28] This impression is fostered in three ways: First, if the word *property* was not customarily used for what belongs to a thing's essence, it was sometimes used as an English equivalent for the Latin *proprium*, "proper accident."[29]

Second, although Clarke does not term the causation of space "ema-nation," he does emphasize the causation's necessity, which is enough to bring out its similarity with the causation of a proper accident by its subject. Finally, the claim that space and time are properties of God is advanced in a discussion of why "space and duration are not *hors de Dieu*"; this invites the thought that part of Clarke's reasoning is that since space and duration are properties of God, existing in him, they do not exist outside of him.

Why would Newton regard this impression as unfortunate? To begin with, Newton believes that space is not an accident because it does not inhere in anything. Moreover, it is a standard theological position that there are no accidents in God.[30] More interesting, per-haps, is a problem spotted by Leibniz: if we are in space and time, and space and time are properties of God, it would seem we have a peculiar, quasi-physical relation to God's being: "If infinite space is God's immensity, infinite time will be God's eternity; and therefore we must say, that what is in space, is in God's immensity, and conse-quently in his essence; and what is in time, [is in the eternity of God and] is also in the essence of God. Strange expressions; which plainly show, that the author makes a wrong use of terms" (L.V.44).[31] Clarke did not help matters by conceding Leibniz's point: "This (sec. 44) strange doctrine, is the express assertion of St. Paul, as well as the plain voice of nature and reason" (C.V.36–48). The apparent admis-sion that Newtonian thinking concerning the relation of space and time to God calls for a literal interpretation of Acts 18:27–28 ("for in him we live and move, and in him we exist") was more than Newton was willing to allow. This issue seems to have been very much on his mind while he composed the Des Maizeaux drafts, as is evidenced by the fact that in one version he explicitly compares Clarke's "figura-tive" use of language to the "figurative" use of language in Acts 18:27–28:

> But as the Hebrews called God MAKOM *place* & the Apostle tells us that he is not far from any of us for in him we live & move & have our being, putting *place* by a figure for him that is in all place; and as the scripture generally spake of God by allusions & figures for want of proper language: so in these Letters the words Quality & Property were used only by a figure to signify the boundless extent of Gods [*sic*] existence with respect to his ubiquity & eternity, & that to exist in this manner is proper to him alone. (p. 99; Draft D)

Newton's primary purpose, then, in composing the Des Maizeaux paragraph is to cancel any impression that space and time are necessary accidents that inhere in God. This explains his remark that "the terms *Quality* & *Property* are not taken in that sense wherein they are vulgarly, by the writers of *Logic* & *Metaphysics* applied to *matter.*" Since a subject is the material cause of its accidents, what Newton is saying here is that Clarke's calling space and time properties of God should not be taken to suggest that God is the material cause of space and time or the subject in which they exist—that is, that space and time are accidents inhering in God.

But Newton does not stop here. If he is not going to retract Clarke's use of *property* outright and admit error, which is out of the question in this polemical setting, then he must make good on his promise to "the reader" to provide a sense of *property* in which space and time can be said to be properties. Pressed by such necessity, he falls back on his conception of space and time as affections of being in general.[32] (Notice that, in so doing, Newton quietly shifts the topic of discussion from the real beings space and time, which is what Leibniz and Clarke were arguing about, to the affections spatial and temporal locatability. Newton's first word on the ontological status of the real beings space and time seems to have been also his last—namely, that space and time have their own manner of existence which fits neither substance nor accident.)[33] Space and time are to be called properties "in such a sense as only implies them to be modes of existence in all beings, & unbounded *modes* & consequences of the existence of a substance which is really necessarily & substantially Omnipresent & Eternal; Which existence is neither a substance nor a quality, but the existence of a substance with all its attributes properties & qualities, & yet is so modified by place & duration that those modes cannot be rejected without rejecting the existence" (p. 97, Draft B). Newton claims here that space and time, taken as properties, are modes of existence, and insists further that a thing's existence is not to be confused with its substance and accidents; and (as McGuire points out) what Newton is saying here is connected with his saying elsewhere that space and time are quantities of existence.[34] What is Newton claiming about spatial and temporal locatability in these texts? And why?

It is useful to begin by considering how a standard scholastic conception of spatial location ill serves Newton's needs. According to Aquinas, both incorporeal and corporeal beings are in space, but in different ways: "Incorporeal things are in place not by the contact of dimensive quantity, as bodies are, but by contact of power."[35] God and the angels lack dimensive quantity and so cannot be in space in the same way that bodies are; but since they can act on places and the bodies that fill them, they are in space "by contact of power." How does Aquinas understand the "dimensive quality" by virtue of which a body is able to be in space? According to Thomas, dimensive quantity is a perfection, intimately linked to membership in the (substantival) genus *corporeal*:

> Body is said to be in the genus substance inasmuch as it has a nature such that three dimensions can be designated in it. Indeed, these three designated dimensions themselves are body according as it is in the genus quantity.
>
> Moreover, it happens in things that what has one perfection may also possess a further perfection. This is evident in man, since he has both a sensitive nature and, beyond that an intellectual nature. Similarly, to this perfection of having a form such that three dimensions can be designated in it, can be added another perfection, such as life or the like.[36]

Newton explicitly opposes this account of divine location; according to him, God is not only omnipresent "virtually," that is, by reason of his power, but also "substantially."[37] One may conjecture that, as spatial and temporal order assumed an increasingly prominent role in the conception of the universe unfolding with the development of seventeenth-century mechanistic science, Newton felt that the Scholastic theory of location failed to recognize the fundamental character of spatial and temporal structure. The Scholastic theory makes the existence of a spatial-temporal order contingent on the creation of certain things (for example, what would have happened, on Aquinas' theory, if God had decided to create only angels?); it also makes the connection of certain things to the nexus of space and time accidental to those things, depending on where and when they happen to exercise their power. One encounters a similar attitude in Leibniz: he sees space as "the order of existence of things possible at the same time" and time as "the order of existence of things possible successively," which is to make spatial and temporal locatability a general feature of created being; accordingly, he holds that all created substances must have bodies.[38] But Leibniz does not make spatial and temporal locatability a condition of all being, making an exception for God who is allowed an existence outside of space and time, whereas Newton does make locatability a condition of all being, holding that even God "is omnipresent not *virtually* only, but also *substantially*." But how is Newton able to do this without giving God dimensive quantity and with it a body?

It is in this dialectical context that Newton's doctrine that space and time are quantities of existence and his related remarks concerning the difference between existence and substance assume their significance. To begin with the latter, to claim that a property such as unity is not substance or accident has a precise technical meaning in Aristotelianism: since real being is divided into substance and accident, to claim that unity is not substance or accident is to claim that it is not a real being or a perfection. When Aquinas takes up, for example, the question "Whether One Adds Anything to Being?" he responds in the negative, as follows: "*One* does not add any reality to *being*; but it is only a negation of division; for *one* means undivided *being*. This is the very reason why *one* is the same as *being*."[39] The fact that Socrates' unity does not add any reality or perfection to the substance and accidents that constitute Socrates' being but is, as it were, presupposed in his being a being at all means, in particular, that his unity is neither a substance nor an accident. It means, paraphrasing Newton, that Socrates' unity is neither a substance nor a quality (that is, an accident); but it is the unity of a substance with all its attributes (essence), properties (*propria*), and qualities (accidents). This is not, of course, to suggest that Socrates' unity is somehow really distinct from his being, from his substance and accidents. Quite the contrary. Since unity holds the peculiar place it does in Aristotelian thought precisely because of its intimate connection with being, one would suppose that Socrates' unity is intimately connected with his being—that is, with his substance and accidents. The point that Newton is making about existence, then, is simply that existence is not a perfection, adding to a thing's reality—that is, that it is not substance or accident.[40]

Now, when Newton makes this observation about existence and links spatial and temporal locatability with existence, he is reworking the Scholastic theory of location. For Aquinas, something owes its (nonvirtual) presence in space to its "perfection of having a form such that three dimensions can be designated in it," a perfection that results from its belonging to genus *corporeal*—that is, from its being a body. For Newton, by way of contrast, being in space does not depend on the possession of the perfection of being a body—or on the possession of any other perfection, for that matter. Being in space, as a general condition of being, like being one, does not result from being *anything* but rather from simply being. Moreover, although Newton might be willing to embrace the Scholastic thesis that a thing's being in space depends on its having "dimensive quantity," what would be quantified, in his view, is not the thing's perfection of being corporeal but rather the thing's existence itself, or better still, "the existence of a substance with all its attributes properties & qualities." Finally, if spatial and temporal locatability are general conditions of a thing's being, and as such independent of whatever reality and perfection the thing happens to have, then Newton can claim, as he does in "Tempus et Locus," "To exist in time and place does not argue imperfection, since it is the common nature of all things" (p. 117). In short, Newton's understanding of spatial and temporal locatability as quantities of existence enables him to place God substantially in space without making God corporeal or implying that he is imperfect.

Newton's revision of the Scholastic theory of location does not bespeak any fissures in the divine being: the theory applied to God simply says that divine spatial and temporal locatability (that is, God's being everywhere and eternal) ought to be viewed along the lines of divine unity and not along the lines of divine omniscience (which can be considered as God's possession, although in an incomprehensibly higher form, of a perfection found in creation). Aquinas, who does not regard unity as a perfection, still holds that unity is an attribute of God, identical with his essence and his other attributes. There is no reason to suppose Newton would have disagreed about God's substantial presence in space and time.

Newton's theory of space and time and their relation to God is reasonably coherent within limits. Still, there are difficulties outstanding. I would like to mention one in closing. Leibniz objects in his last letter to Clarke: "If the reality of space and time, is necessary to the immensity and eternity of God; if God must be in space; if being in space, is a property of God; he will in some measure, depend upon time and space, and stand in need of them. For I have prevented that subterfuge, that space and time are [in God and like] properties of God" (L.V.50). The part of Clarke's fifth letter that most bears on this objection is the following: "God does not exist [sec. 45] in space and in time; but his existence causes space and time. And when, according to the analogy of vulgar speech, we say that he exists in all space and in all time; the words mean only that he is omnipresent and eternal, that is, that boundless space and time are necessary consequences of his existence; and not, that space and time are beings distinct from him, and IN which he exists" (C.V.36–48).

Clarke's claim that space and time are not distinct from God and that God is not in space and time except insofar as the latter are necessary consequences of his existence is of a piece with space and time being *propria* of God. But if space and time are simply God's *propria*, then Leibniz's objection that God depends on space and time and must be in space and time is wide of the mark. It is not obvious, however, that Newton himself can so readily handle Leibniz's worry. As we have seen, Newton retracts Clarke's use of the word *property*.

And he never, to my knowledge, denies, as does Clarke, that space and time are distinct from God. Moreover, Clarke's explanation that when one says that God is in space and time "the words mean only that he is omnipresent and eternal, that is, that boundless space and time are necessary consequences of his existence" seems at odds with the position taken by Newton in "De Gravitatione." To be sure, Newton holds there that space is an emanative effect of God, but his argument for this claim rests squarely on the fact that God, as a being, cannot exist without being related to space in some way: "No being exists or can exist which is not related to space in some way. God is everywhere, created minds are somewhere, and body is in the space that it occupies; and what is neither everywhere nor anywhere does not exist. And *hence it follows* that space is an emanative effect arising from the first existence of being, because when any being is posited, space is posited" ("De Gravitatione," p. 136; my emphasis).[41] It is hard to read this passage without concluding that God's existence in some way depends on space (and time) and that Newton takes God to be in space and time in a stronger sense than does Clarke. To be sure, Newton can point out in response to Leibniz's objection that space and time depend on God since they are his emanative effects. This does not seem to be enough of a response, however. Even if God is able to produce everything that he requires in order to exist (and the coherence of this mutual dependence is open to question), the idea that God's existence depends on real beings that, even if caused by him, are nonetheless distinct from him constitutes a rather striking departure from a standard Western conception of a completely self-sufficient deity.[42]

Notes

1. "De Gravitatione et Aequipondio Fluidorum," in A. Rupert Hall and Marie Boas Hall, *Unpublished Scientific Papers of Isaac Newton* (Cambridge: Cambridge University Press, 1962), pp. 131–132. Henceforth cited as "De Gravitatione." Newton uses the terms *extension* and *space* interchangeably, as can be seen by comparing his discussions on p. 132 and p. 136.

2. "De Gravitatione," pp. 131–132. This is apparently why Newton thinks that space is "more thing-like than an accident, and more closely approaches the nature of substance." We might observe a point of similarity and a point of disagreement between Newton and Leibniz on this score. Like Newton, Leibniz sees a close connection between substantiality and activity, but unlike Newton, he would take the absence of space's activity to be bound up with its ideality and so to remove space not simply from the realm of the substantial but from the realm of the real altogether.

3. See, e.g., "De Gravitatione," pp. 132 and 136. I follow McGuire's translation of *affectio* as "affection" instead of "disposition."

4. Included in this volume. Henceforth cited as PPE.

5. *Annals of Science* 35 (1978), 463–508. Henceforth cited as EAN.

6. McGuire does say that space is "fully actual," but it is unclear what he means by this: Is Homer's blindness, for example, "fully actual" in Homer?

7. Newton does write that space is "non absolute per se" ("De Gravitatione," p. 99), but this should be kept separate from the question of whether space exists in itself or in another.

8. Provided, of course, that space and time are not God.

9. It should be noted that McGuire is not satisfied with Newton's position, as he interprets it: "If God's existence, as well as that of infinite space and time are coeternal, and if the latter are uncreated as Newton claims, how precisely is their existence related to God's? Since they are said to be 'unbounded consequences' of Divine existence, and distinct from God himself, on Newton's theory of predication they are not entailed by Divine nature. Nor are they the result of any sort of creative activity on the part of God. On the other hand, given Newton's conception of what it is for God to exist, it makes no sense that infinite space and time could exist without God's existence. But simply to say that infinite space and time exist because of the sheer existence of an eternal God, is not in itself sufficiently explanatory. Nor, for that matter, is it clear what kind of explanation is being invoked. So to claim that space and time are coeternal with God's existence does not in itself prevent someone from questioning the nature of their relation to Divine existence" (EAN, p. 484).

10. See Julius Weinberg, *A Short History of Medieval Philosophy* (Princeton: Princeton University Press, 1964), p. 116.

11. Maimonides emphasizes this point in chapter 21 of Part II of the *Guide of the Perplexed*, where he compares the relation of an emanative effect to its cause to that of the secondary qualities of a body to its primary qualities: "It is clear that when Aristotle says that the first Intelligence necessarily follows from God, that the second necessarily follows from the first, and the third from the second . . . he does not mean that one thing was first in existence and then out of it came the second as a necessary result . . . By the expression 'it necessarily follows' he merely refers to the causal relation; he means to say that the first Intelligence is the cause of the existence of the second, the second of the third, and so on . . . ; but none of these things preceded another, or has been in existence, according to him, without the existence of that other. It is as if one should say, for example, that from the primary qualities there follow by necessity roughness, smoothness, hardness, softness, porosity, and solidity, in which case no person would doubt that though . . . the secondary qualities follow necessarily from the four primary qualities, it is impossible that there should exist a body which, having the primary qualities, should be denuded of the secondary ones." Cited by Harry Austryn Wolfson, in *The Philosophy of Spinoza* (1934; reprint New York: Meridian Books, 1960), I, 374. Wolfson generally follows Friedländer's translation.

12. *Summa Theologiae*, Part I, Question 77, Art. 6. Henceforth abbreviated as *ST* and cited by part (roman numeral), question (Q.), and article (A.).

13. Newton does write, "This is manifest from the spaces beyond the world, which we must suppose to exist (since we imagine the world to be finite)" ("De Gravitatione," p. 138), but it is unclear what consequences should be drawn from this, since he writes elsewhere in "De Gravitatione" concerning our inability to imagine the infinite: "If anyone now objects that we cannot imagine that there is infinite extension, I agree. But at the same time I contend that we can understand it. We can imagine a greater extension, and then a greater one, but we understand that there exists a greater extension than any we can imagine. And here, inciden-

tally, the faculty of understanding is clearly distinguished from imagination"
(p. 134).

14. Henry More, *Immortality of the Soul*, in *A Collection of Several Philosophical Writings of Dr. Henry More* (1662; reprint, New York: Garland, 1978), p. 28.

15. McGuire, if I am not mistaken, assumes that all (genuine) activity and production require a temporally extended process and so the temporal priority of cause to effect. Such an assumption would lend force to his rhetorical question "In any event, what would it mean to say that God eternally actualizes [infinite space and time]?" (EAN, p. 482). Such an assumption would also explain the rather mysterious transition from his characterization of More's views: "And 'an emanative cause' is such a cause as merely by Being, no other activity or causality interposed, *produces* an effect" (PPE, section ii; my emphasis), to the observation, still associated with More's views, that "but since the notion of an eternal and efficient cause does not involve any activity, *production*, creation, or active efficacy, between it and its effect" (PPE, p. 27; my emphasis): McGuire's reasoning seems to be that, since (genuine) production involves temporal priority ("Causal priority among eternal things does not require the *creation* or *production* of the effect by the cause," PPE, section ii; McGuire's emphasis), More cannot really mean it when he writes that an emanative cause produces an effect. McGuire's assumption that genuine causal activity and production involve a temporally extended process, and, accordingly, the temporal priority of the cause, is questionable, however, especially in the context of this period: Maimonides offers the dependence of a body's secondary qualities on its primary qualities as an example of an instantaneous causal dependence (see n. 11 above), and Aquinas uses the dependence of color on light for a similar purpose. Similarly, the dependence of the universe on God that Descartes outlines in the Third Meditation, where God continually "recreates" the universe at each moment, involves a kind of production or creation that cannot be understood as taking place in time.

16. "Spatiality" is another possibility; but since spatiality is easily confused with extension, I have chosen "spatial locatability" instead.

17. I am unsure what McGuire's attitude would be toward the distinction. On the one hand, he writes, for example, that "accordingly, [Newton] urges that the basic sense of existence is to occupy a definite position, or successively different positions, in space and time. To say of something that it actually exists is to say it is *there* in the objective order of space and time" (EAN, p. 478), which, one might think, calls for a distinction between being in the objective order and the objective order itself. On the other hand, he writes: "So space and time, *qua* affections, simply characterize all being, whether it is substance, attribute or accident, and are really distinct from existing things and independent of their natures. Their mode of predication is therefore external, as they are not real properties which attach to the nature of things in virtue of inherence. Space and time are neither *in* a substance in the manner of its extension and shape, nor are they *in* things as parts in the wholes and as a whole contain parts. Furthermore, they are not accidents or modes of existing things; for this allows they exist only when finite things exist" (EAN, p. 473). Here, McGuire runs together considerations about spatial and temporal locatability (i.e., "space and time, *qua* affections") and space and time taken as real existents ("for this allows that [space and time] exist only when finite things exist"), which seem to me better kept separate. In particular, one should keep distinct Newton's claim that space and time exist independently of finite things (and his related claim that space and time are neither substances nor accidents but have their own manner of existence), which concerns the real existents space and time,

and his thesis that space and time are modes of a thing's existence and, as such, are to be in some sense distinguished from a thing's substance and accidents, which concern not the real beings space and time but the affections spatial and temporal locatability.

18. These draft sheets were written for a paragraph that was included in the Des Maizeaux preface and attributed to Clarke ("M. Clarke has asked me to warn his readers . . . "). The drafts are published in A. Koyré and I. B. Cohen, "Newton and the Leibniz-Clarke Correspondence," *Archives internationales d'histoire des sciences* 15 (1962), 96–97. Hereafter cited by draft letter and page number in Koyré and Cohen.

19. McGuire discusses the Des Maizeaux text in EAN, pp. 475ff., and in PPE, section ii.

20. *Philosophiae Naturalis Principia Mathematica* (London: 1st ed. 1687; 2d ed. 1713; 3d ed. 1726). See also critical edition, I. B. Cohen and A. Koyré (Cambridge, Mass.: Harvard University Press, 1972), II, 761; English edition, Florian Cajori, trans., *Principia* (Berkeley: University of California Press, 1934), II, 545. (The General Scholium was not added until the 2d ed.) Henceforth cited as *Principia* and Cajori, respectively.

21. *Principia* II, 761; Cajori II, 545.

22. C.III.3; I have used H. G. Alexander's edition, *The Leibniz-Clarke Correspondence* (New York: Barnes and Noble, 1976), and follow his practice of citing the correspondence by author's last initial, letter number, and section number.

23. *Principia* II, 761; Cajori II, 545. McGuire is puzzled by Newton's use of *constituit* in this passage: "[Space and time] are not real attributes of God's nature. On the contrary they themselves are, as Newton rather oddly puts it, *constituted* by God's eternal and omnipresent existence. This way of phrasing the point is odd because it implies that space and time are 'ingredients' of God's nature. But Newton explicitly states that God is 'not duration and space'; rather they themselves exist because God exists" (EAN, p. 478). But "constitute" is not the only possible English rendering of the Latin *constituere*; one could (and probably should) translate the passage so that God "sets up" or "establishes" duration and space.

24. These two ideas are connected in "De Gravitatione": "No being exists or can exist which is not related to space in some way. God is everywhere, created minds are somewhere, and body is in the space that it occupies; and whatever is neither everywhere nor anywhere does not exist. And hence it follows that space is an emanative effect arising from the first existence of being, because when any is postulated, space is postulated" (p. 136). Space's being a boundless consequence of God's existence is linked to his being everywhere (his presence reaching from infinity to infinity) in that God cannot be everywhere without effecting infinite space. Presumably similar reasoning would be used to forge a link between God's being eternal and his effecting time. (Hall and Hall inadvertently leave out the word *emanative* in their rendering of this text.)

25. McGuire covers the same ground in EAN, but, it seems to me, without as much explication of the notion of an external common affection: "We can now state more fully what Newton intends to convey in calling place and time 'common affections.' As the passage indicates, they are not properties which inhere in things, but rather characteristics pertaining to their existence. For the weight of interpretation falls on the phrases 'duration of existence' and 'amplitude of presence.' As employed in the passage they indicate that time and place specify anything's actual existence, rather than constituting a part of its defining nature. So considered it is probable that Newton has in mind a traditional distinction among affections. Those which characterize things in virtue of their nature, and those that wholly pertain to their sheer existence. In Newton's view, time and place are in the latter category.

Individual things not only endure in time; they are present in the same or different places through time. Because space and time universally pertain to the duration and presence of things, they are common affections of their existence" (EAN, p. 466).

26. Harry Austryn Wolfson, *The Philosophy of Spinoza* (1934; reprint, New York: Meridian Books, 1960), I, 128.

27. See also C.V.45. It is true that Clarke does not explicitly state in IV.10 that the causation in question is efficient, but the absence of any suggestion otherwise makes this the most natural interpretation of his remarks.

28. This impression may not have been entirely inadvertent; Clarke's employment of the word *property* may reflect some genuine differences between his understanding of the relation of space and time to God and Newton's (see my conclusion).

29. Alexander Koyré points out in *From the Closed World to the Infinite Universe* (1957; reprint, Baltimore: Johns Hopkins University Press, 1982), p. 247, n. 15 (p. 302), that in the French translation of the correspondence, which Clarke himself reviewed, "attribute" is used for "property." At any rate, Des Maizeaux uses the French *propriété* in his translation of Newton's emendation: "Cependant, comme les termes de Qualité ou de Propriété, ont d'ordinaire un sens différent de celui dans lequel il les faut prendre ici." Koyré and Cohen, "'Newton and the Leibniz-Clarke Correspondence," p. 95.

30. See, e.g., Wolfson, *Spinoza*, I, 113–114; Aquinas, *ST* I, Q. 3, A. 6.

31. I have tried to present Leibniz's objection in a way that mitigates Koyré's criticism, on p. 265 of *From the Closed World*, that the objection rests on an equivocation between a spatial sense of *in* and an inherence sense of *in*. Since a thing's *propria* are a part of its being, I take it that to be (spatially) in one of A's *propria* is to be (spatially) in A. It is true that Leibniz claims that Clarke's views commit him to our being (spatially) in God's essence, which would not follow from our being in one of his *propria*; but this usage may be connected with the fact that Leibniz insists on using the word *attribute* where Clarke uses the word *property*. See Koyré, *From the Closed World*, p. 247, n. 15 (p. 302).

32. This is the first time, to my knowledge, that Newton makes public his view that spatial and temporal location are common affections; he did not publish "De Gravitatione" and "Tempus et Locus."

33. "De Gravitatione," pp. 131–132.

34. "De Gravitatione," pp. 136–137; "Tempus et Locus," p. 117.

35. *ST* I, Q. 8, A. 2, R. 1; see also I, Q. 52, A. 1.

36. "On Being and Essence," in Robert P. Goodwin, trans., *Selected Writings of St. Thomas Aquinas* (Indianapolis: Bobbs-Merrill, 1978), p. 40.

37. *Principia* II, 762; Cajori II, 545.

38. Letter to Burcher de Volder, June 30, 1704, in C. J. Gehardt, ed., *Die philosophischen Schriften von G. W. Leibniz* (New York: Georg Olms Verlag, 1978), II, 269; in Leibniz, *Philosophical Papers and Letters* (Boston: D. Reidel, 1976), p. 536.

39. *ST* I, Q. 11, A. 1.

40. In PPE McGuire seems to understand common affections in a similar way when he writes: "These phrases ["infinite space or Immensity" and "endless duration or Eternity"] denote transcendental features because they fall under none of the ten Aristotelian categories nor under the fifth of Porphyry's predicables, accident" (PPE, p. 20), and, a little later, "*Ubi* and *Quando* are therefore 'transcendental' predicates in the precise sense that they refer to conditions of existence that every actually existing thing, God included, must satisfy" (PPE, section ii). This marks a sharpening of his earlier account, in EAN, of the way in which Newton takes

existence to be special. Cf., for example: "Existence *per se* is neither manifested as a single property, nor directly exemplified through properties (*propria*) which are appropriate to the nature of a given thing. It is not substance, since anything that is a substance has a nature and an essence which are its appropriate and defining properties respectively. Moreover, a substance *per se* is that in which its properties exist and through which it itself is perceived. Existence does not fit this ontology. It functions in the second-level statements which refer to the sheer actuality of individual natures. Such statements point to the fact that things exist extra-mentally. As such they exist *qua* actual beings together with their individuating features. Existence is therefore not that which is truly predicated in the category of property: that is, it is not a separate property or attribute which contributes to the defining nature of a thing. Rather, it is an irreducible "feature" of all individuals in so far as they are actual. But to be actual does not confer a distinct property or attribute" (EAN, p. 476).

McGuire subsequently qualifies this account. He suggests that since predications of the form "Socrates exists" are true and convey genuine information about things, it may be that Newton does not deny that existence is a predicate, but rather holds it to be a transcendental predicate: "As it stands, the text does not allow us to state categorically that Newton either affirms or denies that existence is a predicate. He certainly denies that it is a real attribute, property or quality of anything. However, to say this is perfectly compatible with saying that existence is a transcendental predicate. Or put otherwise, it is a second-level predicate: one which functions differently from the attributes of the type that pertain to the defining natures of things" (EAN, p. 477). The clarification in PPE, although welcome, bespeaks a major problem for McGuire's overall interpretation: the closer spatial and temporal locatability and existence become to traditional transcendental properties, such as unity, the less plausible it is to find some real distinction between these properties and their possessors.

41. Although *emanativus* is found in the Latin text reproduced by Hall and Hall (p. 103), it is missing from their translation (p. 136).

42. I would like to thank Thomas Ricketts for his helpful comments.

Chapter 8

Newton's Corpuscular Query and Experimental Philosophy

Peter Achinstein

The most controversial part of Newton's *Opticks* is the set of queries in Book III. Here Newton introduces numerous unproved "hypotheses." This procedure seems strikingly incompatible with his methodological views about empirical science, or "experimental philosophy" (as he calls it), as well as with his actual practice in earlier parts of the *Opticks* and in the *Principia*. One such hypothesis—on which I shall concentrate in what follows—is the subject of Query 29. It states that light consists of particles. More precisely, it asks the question: "Are not the Rays of Light very small Bodies emitted from Shining Substances?"[1] Following this question there is a substantial discussion of the particle theory. And immediately prior to this query there is a discussion for several pages of the rival wave theory, or more precisely, of the hypothesis that "light is supposed to consist in Pression or Motion, propagated through a fluid Medium."

In the present paper I want to consider various interpretations of Newton's procedures in those queries in the *Opticks* that pertain to the corpuscular hypothesis. My aim is to see whether, and if so to what extent, they violate his fundamental ideas about experimental philosophy. To do so some review of these ideas will be necessary. Later I will offer an assessment of part of the Newtonian methodology.

PART I

Newton's Ideas about Hypotheses and Experimental Philosophy

My aim here is to focus on Newton's own views about the proper procedures in experimental philosophy, and to offer some examples in his work where he seems to be following such procedures. (It is not my claim that he always does.) Newton's methodological views have been widely discussed, and some of what I shall say is not new, although I will be making certain claims that I have not found in the literature.

I begin not with the *Opticks* but with a celebrated methodological passage at the end of Newton's *Principia*, immediately following "hypotheses non fingo":

> Whatever is not deduced from the phenomena is to be called an hypothesis; and hypotheses, whether metaphysical or physical, whether of occult qualities or mechanical, have no place in experimental philosophy. In this philosophy particular propositions are inferred from the phenomena, and afterwards rendered general by induction. Thus it was that the impenetrability, the mobility, and impulsive force of bodies, and the laws of motion and of gravitation, were discovered.[2]

Again, in a letter to Cotes in 1713, Newton writes:

> As in geometry, the word "hypothesis" is not taken in so large a sense as to include the axioms and postulates; so in experimental philosophy, it is not to be taken in so large a sense as to include the first principles or axioms, which I call the laws of motion. These principles are deduced from phenomena and made general by induction, which is the highest evidence a proposition can have in this philosophy. And the word "hypothesis" is here used by me to signify only such a proposition as is not a phenomenon nor deduced from any phenomena, but assumed or supposed—without any experimental proof.[3]

In these two passages Newton speaks of "deducing" or "inferring" propositions from phenomena and then making them general by induction. In Rule 4 of the "Rules of Reasoning in Philosophy" in Book III of the *Principia* he combines these locutions and speaks of "propositions inferred by general induction from phenomena." In a letter to Oldenburg (July 1672) he omits a reference to induction, saying simply that "the proper method for inquiring after the properties of things is to deduce them from experiments."[4] In the *Opticks* a reference to induction is also omitted and "deduction" and "inference" are replaced by "proof." Newton begins the *Opticks* by writing: "My design in this Book is not to explain the properties of Light by Hypotheses, but to propose and prove them by Reason and Experiments."[5] And the propositions that follow, which are also called theorems, are defended by providing "proof by experiments." At the end of the *Opticks*, however, the term "induction" reappears when Newton writes:

> As in Mathematicks so in Natural Philosophy, the Investigation of difficult Things by the Method of Analysis, ought ever to

precede the Method of Composition. This Analysis consists in making Experiments and Observations, and in drawing general Conclusions from them by Induction, and admitting of no Objections against the Conclusions, but such as are taken from Experiments, or other certain Truths.[6]

Let me try to provide accounts of some of the Newtonian concepts.

i. Phenomena

In the *Principia* immediately following the "Rules of Reasoning in Philosophy" Newton lists six "Phenomena." They concern the orbits of the satellites of Jupiter and Saturn, the five primary planets, and the moon. Newton offers no published definition of the term *phenomenon*. The following definition, intended for the second edition of the *Principia*, was never published:

> Definition 1. Phenomena I call *whatever can be seen and is perceptible* whatever things can be perceived, either things external which become known by the five senses, or things internal which we contemplate in our minds by thinking. As fire is hot and water is wet, *and gold is heavy,* and *sun is light,* I am and I think. All these are sensible things and can be called phenomena in a wider sense; but those things are *properly called* phenomena which can be seen, but I understand the word in a wider sense.[7]

Although Newton here defines *phenomena* as "whatever can be seen and is perceptible" by the external senses or "internally"—which would seem to allow physical objects such as the sun to count as phenomena—the examples he gives in this definition are all facts rather than objects (fire is hot, water is wet). And this is true as well of the six phenomena listed as such in the *Principia*.[8] For example, the first "Phenomenon" is that the satellites of Jupiter describe areas proportional to the times of description, and their periods are proportional to the 3/2 power of their distance from Jupiter. The third "Phenomenon" is that the five primary planets revolve about the sun.

As is suggested by the unpublished definition, Newton regards these facts as "perceptible." But more than this, in the *Principia* he treats the "phenomena" as facts whose existence not only can be *but has been* established by observations. Concerning Phenomenon 1 Newton writes: "This we know from astronomical observations." That phenomena are established by observations does not necessarily preclude the need for inferences to them. Perhaps some of the examples of phenomena he lists in his unpublished definition can be established directly without inference (especially the "internal" ones: I am,

I think). But the six phenomena of the *Principia* are pretty clearly inferred from what is observed. That the five primary planets revolve about the sun (Phenomenon 3) is not determined to be so simply by seeing the planets revolve. Rather Newton infers that they do revolve "from their moon like appearances" (in the case of Mercury and Venus). Similarly Phenomenon 1 is inferred from astronomical data Newton cites (pertaining to the periods of Jupiter's four satellites and the distances of the satellites from Jupiter's center) by using mathematical calculations. Newton does not say what sorts of inferences to the phenomena are allowable. The important point seems to be that whatever the nature of such inferences, they are "from observations" and they are sufficiently strong to establish the facts in question.

There is another respect in which the discussion of the six phenomena in the *Principia* goes beyond the simple unpublished definition. Newton treats the phenomena of the *Principia* as noncontroversial, or potentially so. They are not facts that although observed by some scientists are disputed by others. They are facts that scientists aware of the results of observations that have been made do agree on, or would if they came to be aware of them. In the case of Phenomenon 1 Newton gives the results of observations of three different astronomers which establish this phenomenon. And he writes that Phenomenon 4 (Kepler's third law) "is now received by all astronomers."[9]

In the *Opticks* "phenomena" are not listed as such. Nor does the word appear when Newton proceeds to give a "proof by experiment." Yet there is something here that corresponds to Newton's notion of a phenomenon in the *Principia*—specifically an established fact about the result of the (type of) experiment that anyone who performed the experiment would agree to on the basis of observation. For example, Proposition 2 of Book I of the *Opticks* states that the light of the sun consists of rays differently refrangible. To prove this Newton describes a series of experiments, the first of which involves the passage of sunlight through a single prism and the formation of the solar image on a sheet of paper. This image, writes Newton, is "oblong and not oval [as might otherwise be expected], but terminated with two Rectilinear and Parallel Sides, and two Semicircular Ends." Newton's description of the results of this experiment are, I suggest, descriptions of a "phenomenon." It is a fact—one that Newton has determined to be so by observation and that can be so determined by anyone performing this type of experiment—that an oblong image is produced in such an experiment. This fact would be agreed to by all, even if there were doubts about whether it proves the proposition that light is composed of rays of different refrangibility.

ii. Deducing Propositions from Phenomena

Following the six "Phenomena" in Book III of the *Principia* is a set of what Newton calls propositions (as well as theorems). These propositions are inferred from one or more of the phenomena and from earlier propositions in the *Principia*. In some cases the inference is deductive in a sense that any contemporary philosopher or logician could acknowledge. For example, the first part of Proposition I of Book III states that Jupiter's satellites are continually drawn off from rectilinear motion and retained in their orbits by a centripetal force directed to Jupiter's center. This follows deductively from Phenomenon 1 (given above) and from Proposition II, Book I (which states that every body that moves in any curved line described in a plane, and by a radius drawn to a point either immovable or moving forward with a uniform rectilinear motion, describes about that point areas proportional to the times, is urged by a centripetal force directed to that point).

Accordingly, when Newton speaks of "deducing" propositions from phenomena, he means to include at least ordinary deductions—deductions of a type that would be found in mathematical proofs. However, Newton also includes inductions.[10] I shall follow Ernan McMullin and Maurice Mandelbaum in supposing that by induction Newton has in mind an inference from some property found to hold for all observed members of a class to the claim that this property holds for some members of that class that have not been observed or for all members.[11] The inferences he cites in his rules of reasoning are of this type. Here are two examples:

> The bodies which we handle we find impenetrable, and thence conclude impenetrability to be a universal property of all bodies whatsoever. That all bodies are movable, and endowed with certain powers (which we call the inertia) of persevering in their motion, or in their rest, we only infer from the like properties observed in the bodies which we have seen.[12]

Although Newton does not explicitly use the term *induction* here, he does later in the General Scholium in Book III when referring to such inferences concerning impenetrability, mobility, and so on. (See the first quotation above.)

Does Newton impose any conditions on induction? He imposes none on the number of members of the class that need to be observed. But in his third rule of reasoning he does introduce a condition that might be interpreted as affecting the kinds of properties subject to induction. The third rule says:

> The qualities of bodies, which admit neither intensification nor remission of degrees, and which are found to belong to all bodies within the reach of our experiments, are to be esteemed the universal qualities of all bodies whatsoever.

For present purposes three questions need to be asked. First, what does Newton mean by qualities that "admit neither intensification nor remission of degrees"? Second, is he committed to the view that where bodies are concerned, inductions are possible only where qualities are of this sort? Third, is Rule 3 to be construed as governing all inferences whether or not they involve bodies?

The phrase "admit neither intensification nor remission of degrees" (or, in another translation, "admit neither intension nor remission") is a scholastic one that Newton probably used in his own special way. McGuire suggests that for Newton a quality subject to "intension and remission" is one that "can manifest continuous and successive degrees of intensity: as can the pitch of a sound or the depth of a color." Even more generally, McGuire takes such qualities to be those that admit of "more or less" or of differences of degree.[13] In his discussion following Rule 3 Newton does use the expression "not liable to diminution" to refer to the qualities he has in mind. And his examples of qualities to which the rule applies (and hence presumably qualities that "admit neither intensification nor remission of degrees") are extension, hardness, impenetrability, mobility, and inertia. Newton's discussion contains no examples of qualities that can be "intended and remitted," but in an earlier draft of this rule he does list heat and cold, wet and dry, light and darkness, and color and blackness.[14]

Is Newton committed to the view that, where bodies are concerned, inductions can involve only properties that cannot be "intended and remitted"? This is doubtful. In the discussion following Rule 3 he admits that the *gravity* of bodies "is diminished as they recede from the earth." Accordingly, it should be a property with "intension and remission," and hence not subject to Rule 3. Yet in this passage, as well as in Corollary II of Proposition VI, Newton explicitly says that the property of gravitating toward the earth is subject to Rule 3. Moreover, hardness and impenetrability, which Newton cites in his discussion of Rule 3, might be said to admit of degrees. Yet these properties are supposed to be subject to the rule. McMullin has suggested that perhaps by "hardness" here Newton did not mean the propery that admits of degrees but the property of possessing (some degree of) hardness (impenetrability, gravity, and so on).[15] But then, as McMullin notes, the same could be said for qualities such as color and heat (possessing some degree or intensity of color and heat); yet Newton explicitly classifies color and heat as being subject to "intension and remission." McMullin concludes that Newton "could have omitted the troublesome intensity criterion from the published version of Rule 3, without in the least affecting the manner of applying the Rule to concrete cases."[16] I tend to agree.[17]

Finally, it is even more dubious to suppose that Newton meant that where bodies are not concerned, inductions are restricted to qualities that cannot be "intended and remitted." In the *Opticks* (as we shall see) Newton needs to make inductions involving the property of refrangibility, which by his own admission is subject to degrees. These inductions concern light rays which Newton does not want to have to assume at the outset to be bodies. Accordingly, in what follows I shall not construe Newton as requiring that the properties subject to induction be those that cannot be "intended and remitted."[18]

When Newton speaks of "deductions from phenomena," does he mean to include more than inductions and (ordinary) deductions? In addition to his Rules 3 and 4,[19] which involve induction, there are Rules 1 and 2, which pertain to *causes*. Out of considerations of simplicity, we should admit no more causes of natural things than are true and sufficient to explain the appearances (Rule 1); and therefore, to the same natural effects we should, so far as possible, assign the same causes (Rule 2). As examples of the latter, Newton cites the cause of the descent of stones in Europe and in America, and the cause of the reflection of light in the earth and in the planets. Newton does not relate the first two (causal) rules to the last two (inductive) ones. (He does not claim, for example, that the causal rules are special cases of the inductive ones.) Moreover, he uses all four rules in the early parts of Book III in a series of arguments leading to his law of universal gravitation. Both facts suggest that he thought of his causal rules and his inductive ones as distinct.

What sorts of inferences are Rules 1 and 2 supposed to generate, and how might the difference between these and inductions be expressed? Newton may well have several types of inferences in mind that are sanctioned by his first two rules. Let me note two that he explicitly uses. He begins the argument for Proposition V, Book III, as follows:

> For the revolutions of the circumjovial planets around Jupiter, of the circumsaturnal about Saturn, and of Mercury and Venus, and the other circumsolar planets, about the sun, are appearances of the same sort with the revolutions of the moon about the earth; and therefore, by Rule 2, must be owing to the same sort of causes. (p. 410)

Newton seems to be arguing from facts about the observed Keplerian motions of the moons of Jupiter, Saturn, and so on, described in his six phenomena, to the proposition that these motions have the same causes. The form of the inference is this:

C_1:   Effects $E_1, \ldots, E_n$ are the same in systems $S_1, \ldots, S_k$.
Therefore (by Rule 2) these effects have the same causes in all these systems.

Here, as well as in the examples he gives following the introduction of Rule 2, Newton simply infers that these effects in different systems are produced by the same cause, without identifying that cause. Perhaps he believes that an induction will permit an inference from the fact (phenomenon) that various known planets and their satellites satisfy Kepler's laws to the proposition that all the planets and their satellites do; whereas it will not permit an inference from the similar motions of these bodies to the claim that these motions being similar, their causes are as well.

A second use of Rules 1 and 2 is illustrated by parts of his discussion of Proposition IV. Here these rules are explicitly invoked in sanctioning an inference from the fact that the force keeping the moon in its orbit and the force of gravity on the earth both have features in common—they are both responsible for motions of bodies, they are both centripetal forces, and they are both inverse-square forces—to the claim that they are one and the same (type of) force: gravity. (In saying that they are one and the same force Newton seems to mean that not only do they obey all the same laws but also that were they both to act on a body, the total force exerted would be the same as that exerted by either.)

This second type of argument might be given the following general form:

C_2:   The cause of effects E (e.g., motion of a certain type) in system 1 is an x (a type, e.g., a force) with properties $P_1, \ldots, P_n$.
The cause of the same or similar effects E in system 2 is an x with properties $P_1, \ldots, P_n$.
Therefore (by Rules 1 and 2)
The cause of effects E in system 1 is the same x (e.g., the same force) as that which causes E in system 2.

Here, by contrast with the previous inference, information about the cause is provided.

How is C_2 different from induction? From the first two premises in C_2 induction would allow us to conclude:

The cause of effect E in every similar system (or in some unobserved one) is an x with properties $P_1, \ldots, P_n$.

But, on this interpretation, induction would not allow us to conclude
that the cause is the *same* (type of) x. Thus, from the fact that the force
retaining Mercury and Venus in their orbits around the sun is centripetal and varies inversely as the square of the distance, induction
permits Newton to conclude that the force retaining the other planets
in their orbits is also centripetal and inverse square. But Newton
seems to believe he needs additional rules to infer that these inverse-
square centripetal forces are the same (that is, gravity). He seems to
believe that while induction allows generalizing that a cause with the
features in question will operate in other cases, it will not permit an
inference that we are dealing with one (type of) cause here, not
many.

In addition to C_1 and C_2 there may well be other types of causal
inferences Newton regards as sanctioned by Rules 1 and 2, for
example:

C_3:   Effects $E_1, \ldots, E_n$ in system 1 are caused by C.

Effects $E_1, \ldots, E_n$ are also present in system 2.

Therefore (by Rules 1 and 2) effects $E_1, \ldots, E_n$ are caused by C in
system 2.

Let me call inferences C_1–C_3—and any others Newton would regard
as generated by the first two rules—examples of *causal simplification*.
The reason for choosing this name is that Newton explicitly justifies
Rule 1 (and Rule 2, which he seems to think follows from it) by an
appeal to simplicity ("Nature is pleased with simplicity"). In the introduction to the *Principia,* and especially in an unpublished preface
to the *Opticks,*[20] Newton stresses the idea of deriving everything in
physics from a *few* general principles rather than inventing a new one
for each phenomenon. Perhaps his thinking is that induction will
yield principles that are general but not necessarily few in number. By
induction we may arrive at the generalization that the cause which
keeps each of the planets in its path around the sun is some centripetal inverse-square force. But this is compatible with the idea that
these forces are different in other respects, thus requiring a separate
law for each. To simplify and unify the situation, Newton's Rules 1
and 2 permit him to say that, until new phenomena show otherwise,
from the fact that these forces have the same observed effects, we
may infer that they are one and the same force.[21]

In the case of induction Newton imposes no restrictions on the
numbers of observations, and (arguably) none on properties or

classes. The same is true of causal simplification. He offers no restrictions on the types or numbers of effects that must be observed in order to infer identical causes. What does seem reasonably clear is that Newton treated these principles differently—both in the rules and in the arguments for early propositions in Book III. Moreover, if we are to construe those arguments as being or containing "deductions from the phenomena," then the latter category will include ordinary deductions, inductions, and causal simplification. There are propositions (such as Propositions I and II of Book III) that he derives from the phenomena and previous propositions by (ordinary) deduction alone without inductive generalization or causal simplification. There are also propositions (such as Proposition IV) that he derives using inferences of all three types. (See, in particular, the alternative "demonstration" that Newton offers for Proposition IV in the Scholium, p. 409.)[22]

Accordingly, I suggest that for Newton "deduction from the phenomena" is a certain form of reasoning from facts regarded as established by observation. Such reasoning can be deductive in the ordinary sense; it can be inductive in the sense of inferring properties of all or some members of a class from properties of all the observed members; it can be causal simplification; or it can be a combination of these. (In what follows I shall use "deduction" in quotes, either by itself or in "deduction from phenomena," when speaking of it in Newton's sense, and without quotes when speaking of it in the ordinary sense.)

Two points concerning this definition deserve note. First, Newton regards a proposition "deduced from phenomena" as being established "beyond reasonable doubt" (my expression), or as having the "highest evidence that a proposition can have in this [experimental] philosophy" (Newton's expression). This does not make such a proposition *incorrigible* for Newton. A proposition "deduced from phenomena" may be false, but only further phenomena can show this. Merely imagining a "contrary hypothesis" (in the style of Descartes' method of doubt) is not sufficient to diminish the extent of its believability on the basis of the phenomena. This is clear from Newton's Rule 4 in the *Principia* (see n. 19).

Second, in his discussion of the third rule of reasoning Newton allows inferences from some observed members of a class to other members of the class, even if the latter are not only not observed but unobservable in principle:

> The extension, hardness, impenetrability, mobility, and inertia of the whole result from the extension, hardness, impenetrability, mobility, and inertia of the parts [this presumably Newton

has inferred from observations]; and hence we conclude the least parts of all bodies [which we have not seen and which may not be observable] to be also all extended, and hard and impenetrable, and movable, and endowed with their proper inertia.[23]

This form of reasoning allows Newton to claim that if all observed members of a certain class have a property, then this property can be ascribed to all members of the class, even if some are unobservable. The rule itself, however, does not commit him to postulating unobservable members of the class.[24] In the passage just quoted I take Newton to be saying that extension, hardness, and so on are applicable to the least parts of bodies, if such there be, even if these parts are unobservable in principle. He admits that

> In the particles that remain undivided, our minds are able to distinguish yet lesser parts, as is mathematically demonstrated. But whether the parts so distinguished, and not yet divided, may by the powers of Nature, be actually divided and separated from one another, we cannot certainly determine.[25]

So Rule 3 does not commit Newton to a belief in atoms (that is, physically indivisible parts), but only to the position that if they exist, then they are extended, hard, impenetrable, and so on.

As I have noted, in the *Opticks*, by contrast to the *Principia*, Newton does not write down "phenomena" as such. Instead he describes a series of experiments. Nor does he use the expression "deduction from the phenomena." Instead he speaks of "proof by experiment." Whether or not these expressions had the same meaning in Newton's mind,[26] the "proofs" Newton offers in the *Opticks* do, or can readily be made to, conform to the previously noted criteria of "deductions from phenomena." Consider the first proposition in the *Opticks* and the experiment Newton suggests to prove it.

Proposition I states: "Lights which differ in Colour differ also in Degrees of Refrangibility." In the first experiment he obtains a black, oblong, stiff paper, which he divides into two equal parts, painting one part red and the other blue. He then views the paper through a prism and finds that when the refracting angle of the prism is turned upward, so that the paper seems to be displaced upward by the prism, its blue half is displaced upward farther than its red half. Similarly, when the refracting angle of the prism is turned downward, the blue half is displaced downward farther then the red half. From this Newton draws the following conclusion:

> Wherefore in both cases the Light which comes from the blue half of the Paper through the Prism to the Eye, does in like Circumstances suffer a greater Refraction than the Light

which comes from the Red half, and by consequence is more refrangible.[27]

As a "proof" this is somewhat incomplete. But I suggest that it can be reconstructed in a more complete way, using one of Newton's previous axioms in the *Opticks*, as follows:

(1) We begin with the following result of Newton's first experiment, which we can call a phenomenon, since it is a fact that Newton himself has established by observation and he thinks will readily be accepted by anyone performing the experiment: when the paper, one half of which is red and the other half blue, is viewed through a prism, the blue half is more displaced than the red half.

(2) From Axiom VIII and the discussion following it (pp. 18–19): when an object is seen through a prism, the rays of light from that object are refracted, and the object is seen not in its "proper" position but in some displaced position as a result of this refraction.

(3) From (1) and (2), by (ordinary) deduction, the blue rays coming from the colored paper are more refracted by the prism than the red.

(4) From (3), by induction, blue rays are more refrangible than red—or even more generally, rays of different colors are differently refrangible.

So reconstructed, we have an (ordinary) deduction of the proposition in (3) from the phenomena (plus previous propositions). And in (4) we make this general by induction. The last step is crucial because Newton is obviously generalizing from his experiments with the type of prism he uses to other prisms, indeed to any case in which there is refraction of light from objects painted blue and red, whether or not a prism is involved. Even more generally, in order to arrive at the proposition he is trying to prove he needs to infer that such differences in refrangibility are present in all colors, not just red and blue. Whether the argument is sufficiently strong to establish this conclusion I will not explore here (but see Section vi). My claim is only that if this argument, or something like it, is a reasonable reconstruction of what Newton is doing here, then Newton's first "proof" in the *Opticks* is a perfectly good example of what Newton in the *Principia* and other writings would classify as "deducing propositions from the phenomena and making them general by induction."[28]

It is not my claim that each of Newton's "proofs" in the *Opticks* is successful. In some (perhaps even in the proof just given) the inductive generalization might be considered too sweeping if based only on the experiments mentioned. In others a proposition is introduced as a premise that is not "deduced from the phenomena" in the sense just indicated and is indeed controversial.[29] Whatever the validity or persuasiveness of the "proofs by experiment" in the *Opticks*, I suggest that Newton treated them as "deductions from phenomena." He thought of them as demonstrating the truth of the propositions beyond reasonable doubt from facts themselves established by observing the results of experiments—facts that anyone capable of performing the experiments could agree on.

iii. Hypotheses and Their Rejection

Newton used the word *hypothesis* in different senses in his writings.[30] But there is one use that I shall focus on in what follows, since it is, I think, central for understanding Newton's professed methodology. This is the use of *hypothesis* in the first two quotations at the beginning of this paper.

> Whatever is not deduced from the phenomena is to be called an hypothesis.[31]

> And the word "hypothesis" is here used by me to signify only such a proposition as is not a phenomenon nor deduced from any phenomena, but assumed or supposed without any experimental proof.[32]

A hypothesis is any proposition that is not "deduced from phenomena" in the sense of this expression explicated above. Newton allows that hypotheses (in this sense) may be "metaphysical or physical . . . , of occult qualities or mechanical."[33] Presumably, however, he does not include purely mathematical propositions, even though they are not "deduced from phenomena."[34] When a proposition is "deduced" from phenomena, it has "the highest evidence that a proposition can have in this [experimental] philosophy."[35] Hypotheses do not have such evidence. Whether Newton thought that hypotheses can have some evidence in their favor, just not the "highest," or whether he meant by a hypothesis a proposition for which no evidence at all has been proposed is not perfectly clear. (The second quotation might suggest the latter interpretation, except that Newton here speaks of "experimental proof" rather than "evidence"). In what follows I will take *hypothesis* in the broader sense in which a proposition is a hypothesis if it is not "deduced" from phenomena, even if there is some evidence in its favor. Accordingly, a hypothesis does not have the highest evidence achievable in experimental philosophy; it is not established beyond reasonable doubt, although there may be some reason to believe it.

One potential ambiguity must be cleared up. When Newton says that a hypothesis is a proposition that is "not deduced from phenomena," does he mean (a) that it has not actually been derived (in the way I have suggested), or (b) that it is not deriv*able* from phenomena, or perhaps even (c) that it is not derivable from phenomena or from any facts that will become phenomena? I shall understand him to be asserting (a), which is of course entailed by both (b) and (c), though not conversely. Newton's primary concern in experimental philosophy is to provide reasons for believing propositions that are as strong as possible in this philosophy. Such reasons have not been provided for a proposition that is deducible from phenomena but not yet deduced. Propositions of this sort do not have the highest evidence possible. Accordingly, a proposition might be a hypothesis at one time but not at another. Further observations and experiments could change its status.[36]

Now in some sense or other Newton professes to reject hypotheses, though in what sense is not perfectly clear. For example, in the last paragraph of the *Principia*, which immediately follows the paragraph in which he writes "hypotheses non fingo," Newton introduces what is clearly a hypothesis "concerning a certain most subtle spirit which pervades and lies hid in all gross bodies; by the force and action of which spirit the particles of bodies attract one another at near distances, and cohere if contiguous."[37] Newton seems to recognize that this is a hypothesis since he writes that we are not "furnished with that sufficiency of experiments which is required to an accurate determination and demonstration of the laws by which this electric and elastic spirit operates."[38] How, then, can Newton be said to reject hypotheses?

One thing he clearly does reject is a certain version of the so-called method of hypothesis. In a letter to Oldenburg of June 2, 1672, he writes:

> If any one offers conjectures about the truth of things from the mere possibility of hypotheses, I do not see how anything certain can be determined in any science; for it is always possible to contrive hypotheses, one after another, which are found rich in new tribulations.[39]

Newton rejects inferences to the truth of a hypothesis from its "mere possibility"—that is, from its ability to explain some observed phenomena and/or from the fact that it entails these phenomena. I take Newton to be rejecting inference forms such as these:

Phenomena $p_1, \ldots, p_n$ are established by observation.
Hypothesis h, if true, would explain $p_1, \ldots, p_n$.
Therefore h is true.

Phenomena $p_1, \ldots, p_n$ are established by observation.
$p_1, \ldots, p_n$ are derivable from hypothesis h.
Therefore h is true.

Newton's objection is that numerous hypotheses can be contrived on the basis of which the phenomena can be derived and explained. Such forms of reasoning do not establish the truth of hypotheses. They do not confer upon hypotheses the highest certainty possible in experimental philosophy. Accordingly, such inferences are rejected in this philosophy.

Even if this is clearly a part of Newton's position on hypotheses, the rest of that position is not so clear. Here are two possibilities:

(A) In experimental philosophy one can introduce hypotheses for various purposes, *so long as one does not infer their truth.* For example, in the passage from which the last quotation was taken, Newton allows one to construct explanations of observations by means of hypotheses, and also to use hypotheses "as an aid to experiments" (perhaps to suggest new experiments). But from the fact that the hypothesis does explain the observations and even predicts the results of new experiments, one cannot infer its truth. Such an inference will not yield the highest certainty possible in experimental philosophy; in that philosophy only inferences that do yield such certainty—only "deductions from phenomena"—are permitted.

(B) In experimental philosophy one should avoid not only inferring the truth of hypotheses but also introducing them for any purposes. In the same passage in which Newton seems to allow hypotheses to be introduced to explain observations and to serve as "an aid to experiments" he also writes: "Wherefore I judged that one should abstain from considering hypotheses as from a fallacious argument."[40] And, of course, in the *Principia* itself Newton writes, "Hypotheses . . . have no place in experimental philosophy."[41]

Newton's *pronouncements* about hypotheses in experimental philosophy do not clearly favor one or the other of these interpretations. In his actual practice, as we have already noted, he in fact does introduce (what even he would regard as) hypotheses. So perhaps (A) reflects his position more accurately than (B). There is another possibility, however, as follows.

(C) The highest aim in experimental philosophy is to establish propositions with certainty, or at least with as much certainty as is possible for a science that is based on experience. The greatest certainty possible in experimental philosophy is achieved when propositions are "deduced from phenomena." Since, by definition, hypotheses are not "deduced from phenomena," they do not have the greatest certainty possible in empirical science. Accordingly, the highest aim here is not achieved by inferences to hypotheses. If there is some proposition of interest to us that we think is or might be true and it is possible to "deduce" it from the phenomena, or if it is possible to construct new experiments that will yield new phenomena from which it can be "deduced," then the "deduction" should be made. There may, however, be some proposition of interest to us for which we have found no "deduction from phenomena." In such a case we are not entitled to infer that it is true on the grounds that it explains and/or predicts phenomena. We are entitled to consider it, however, to determine what sorts of explanations and predictions it yields. Moreover, we may be able to give some reasons, some evidence, in its favor, even if this is not the highest possible evidence. If we can give such reasons, we should. (What these might be will be discussed in Part II, where I turn to Queries 28 and 29 in the *Opticks*.) We must recognize, however, that whatever their nature, these reasons are not the strongest, and we should continue to search for better ones.

Indeed, the last clause in the previous paragraph may reflect another sense in which Newton means to be rejecting hypotheses. He is rejecting the idea (which he may be attributing to Hobbes and Descartes) that once one has introduced a hypothesis (or a set of them) to explain the phenomena and shown that the phenomena are consistent with the hypothesis, *there is nothing more to do.*[42] On the contrary, although it can be perfectly legitimate to introduce hypotheses to explain phenomena, this is only the beginning. One must try to "deduce" the hypothesis from these or other phenomena. One must try to rid the hypothesis of its hypothetical nature.

Newton's pronouncements about hypotheses do not form a consistent set, since both (A) and (E) can be found in them. If one also takes into account Newton's actual practice, however, then perhaps his overall position on hypotheses is best represented by (C). Admittedly, (C) is incompatible with some of his pronouncements (for example, "hypotheses . . . have no place in experimental philosophy"), as well as with some of his practice (for example, in the *Opticks* the introduction of the supposition that a force acts on light rays to produce refraction—a supposition not "deduced from phenomena" but treated with the certainty of something that is. See n. 29). If, however, we want to attribute a consistent view to Newton that reflects a good deal, though by no means all, of his pronouncements and his practice, then I think that (C) is superior to (A) or (B).[43]

Interpretation (C) allows us to view the Queries in the *Opticks* in a way that does not violate Newtonian methodology. Indeed, as I will show, (C) is suggested by some remarks Newton makes about his methodology in Book III of the *Opticks*. Finally, as I have noted, (C) admits the possibility of providing reasons for hypotheses, even though these do not generate the certainty of "deductions from phenomena." There are occasions in Newton's actual practice when he provides such reasons, *and explicitly recognizes this.* Thus, in a letter to

Oldenburg in July 1672, Newton responds to a criticism by Hooke who accuses Newton of assuming as a hypothesis that light is composed of bodies. He writes:

> It is true that from my theory I argue the corporeity of light, but I do it without any absolute positiveness, as the word *perhaps* intimates, and make at most but a very plausible consequence of the doctrine, and not a fundamental supposition, nor so much as any part of it which was wholly comprehended in the precedent propositions.[44]

Again in a letter to Boyle of February 1678, after proposing various hypotheses about the aether (including one concerning the cause of gravity), Newton writes: "By what has been said, you will easily discern whether, in these conjectures, there be any degree of probability; which is all I aim at."[45]

I believe that Newton was proposing such nonconclusive probabilistic reasons for at least some of the hypotheses in the Queries; in Part II I will attempt to analyze such reasoning. Newton says that "deduction from the phenomena" is the highest evidence that a proposition can have in experimental philosophy; he does not say that it is the *only* evidence.

iv. Analysis and Synthesis
At the end of the *Opticks* Newton offers the following methodological remarks:

> As in Mathematicks, so in Natural Philosophy, the Investigation of difficult Things by the Method of Analysis, ought ever to precede the Method of Composition. This Analysis consists in making Experiments and Observations, and in drawing general Conclusions from them by Induction, and admitting of no Objections against the Conclusions, but such as are taken from Experiments, or other certain Truths. For Hypotheses are not to be regarded in experimental Philosophy. And although the arguing from Experiments and Observations by Induction be no Demonstration of general Conclusions; yet it is the best way of arguing which the Nature of Things admits of, and may be looked upon as so much the stronger, by how much the Induction is more general . . . By this way of Analysis we may proceed from Compounds to Ingredients, and from Motions to the Forces producing them; and in general, from Effects to their Causes, and from particular Causes to more general ones, till the Argument end in the most general. This is the Method of Analysis: And the Synthesis consists in assuming the Causes discover'd, and estab-

lish'd as Principles, and by them explaining the Phenomena proceeding from them, and proving the Explanations.[46]

Although Newton does not use the phrase "deduction from phenomena" in this passage, I take it that in "Analysis" one provides such "deductions." From the results of experiments and observations one proceeds to draw general conclusions by induction; and "it is the best way of arguing which the Nature of Things admits of." But Newton also speaks here of the method of composition or synthesis. His idea seems to be that once we have "deduced" some proposition from the phenomena, we can then use that proposition in explaining phenomena "proceeding from" it. He also speaks of "proving the Explanations," although he does not say what this means or how it is to be done.[47]

Following this general methodological passage, Newton writes that in the first two books of the *Opticks* he has proceeded by means of "Analysis to discover and prove the original Differences of the Rays of Light in respect of Refrangibility, Reflexibility, and Colour. And these Discoveries being proved may be assumed in the Method of Composition for explaining the Phenomena arising from them." Newton has something different to say about his procedure in the third book however:

> In this third Book I have only begun the Analysis of what remains to be discover'd about Light and its Effects upon the Frame of Nature, hinting several things about it, and leaving the Hints to be examin'd and improv'd by the farther Experiments and Observations of such as are inquisitive.[48]

Newton seems to be referring here to the Queries in Book III rather than to the eleven "Observations" with which the third book begins.[49] But these Queries contain propositions that are not deduced from phenomena. The "Analysis" has only begun, and further experiments and observations will be necessary to complete it. Nevertheless, Newton obviously believes it is of value to introduce such hypotheses. And what he has done (the "hints" he has given) may provide some reasons, albeit not conclusive ones, for believing the hypotheses introduced. This conforms with interpretation (C) above.

PART II

Light

v. The Corpuscular Hypothesis of Light
The final sixty-seven pages of the *Opticks* contain a set of thirty-one Queries, most of which are followed by discussions, some fairly ex-

tensive. What will be of particular concern here is Query 29, in which Newton asks whether rays of light are not very small bodies emitted from shining substances. There follows a discussion in which Newton points out that such very small bodies will have various observed features of light. For example, they will pass through a uniform medium in straight lines without bending into the shadow. They will be subject to certain forces that will permit them to satisfy the laws of reflection and refraction. Their different sizes will produce the different observed colors and degrees of refrangibility. By the forces they exert they can cause vibrations which produce fits of easy reflection and transmission observable in the phenomenon known as Newton's rings. And so forth.

Whatever Newton is supposing about the proposition that light consists of small bodies or particles (and the ancillary proposition that these particles can exert, and be acted upon by, forces subject to his laws of motion), he is not supposing that this proposition is "deduced from phenomena." Accordingly, it is a *hypothesis,* and one that he invites us to consider seriously. In doing so is he violating his methodology? Yes, if that methodology is to be construed in accordance with interpretation (B) of Section iii, which forbids the introduction of hypotheses in experimental philosophy for any purposes. But (B) is a fairly radical interpretation of Newton's methodology. What about (A) and (C)? He would be violating both of these if from the fact that the particle hypothesis, if true, can explain a variety of phenomena he is inferring the truth of that hypothesis. For then he would be employing a version of the method of hypothesis that he rejects and that both (A) and (B) preclude. Newton, however, does not, at least explicitly, claim that the particle hypothesis is true. And before we saddle him with such a view, let us consider alternatives.

One is that all he is doing is *considering* the hypothesis—seeing what follows from it and what it can explain—and not claiming that the fact that it can explain such phenomena provides a reason to believe it true. In short, the query is to be construed as just that—a question—which makes no commitment to the hypothesis in the question. Such an interpretation seems to be the one offered by A. I. Sabra when he writes:

> These are *hypotheses* advanced by Newton without consideration of their truth or falsity. The fact that they appear in the *Opticks* as Queries and not as Propositions means that they do not form part of the *asserted* doctrine of light and refraction.[50]

This interpretation is compatible with both (A) and (C), which allow one to introduce hypotheses for various purposes, so long as one does not infer that they are true.

There is another interpretation, one particularly suggested by (C). On this interpretation Newton is doing more than simply considering the hypothesis by seeing what follows from it and what it can explain. He is providing some grounds for believing it is true, or at least he is attempting to do so, although not the "highest evidence" possible. What grounds are these?

Is he perhaps saying that the fact that the particle hypothesis (together with ancillary assumptions) *can explain a variety of observed optical phenomena* provides some grounds for believing that hypothesis? That is, is he accepting an inference of the following type:

> Phenomena $p_1, \ldots, p_n$ are established by observation.
>
> Hypothesis h, if true, would correctly explain $p_1, \ldots, p_n$ (or $p_1, \ldots, p_n$ are deducible from h).
>
> Therefore (probably) h.

Such an inference is a version of the method of hypothesis, though not the one rejected earlier. In the present case, as opposed to the previous one, h is inferred only with probability. On this interpretation Newton would allow us to use the method of hypothesis as long as we recognize that the conclusion is not drawn with certainty (or with the highest certainty possible in experimental philosophy). This interpretation is compatible with both (A) and (C), since h here is inferred only with probability, not certainty. Indeed, (C) speaks explicitly of a form of reasoning to a hypothesis which provides some reason for believing it, but not the highest evidence possible.

Let us call an inference to a proposition that furnishes that proposition with the highest evidence possible in experimental philosophy a *strong* inference. Let us call an inference to a proposition that provides some evidence for that proposition, but not the highest possible in experimental philosophy, a *weak* inference. For Newton, in experimental philosophy all and only strong inferences are "deductions from phenomena." Now, on the present interpretation, Newton is proposing a weak inference to the hypothesis that light rays are corpuscular.

If so, it is dubious that the inference just stated is what he has in mind. For one thing, the fundamental objection he raises against the earlier version of the method of hypothesis is applicable to this one as well. The objection is that numerous hypotheses can be devised that will explain the same phenomena. Having done so, we cannot infer each of these hypotheses with probability.

Second, to say that Newton's inference to the corpuscular hypothesis involves reasoning only of the type just stated would be to ignore completely his discussion in Query 28, which immediately precedes the query in which the corpuscular hypothesis is introduced. In Query 28 Newton considers the wave theory, which he takes to contain "all Hypotheses . . . in which Light is supposed to consist in Pression or Motion, propagated through a fluid Medium" (p. 362). In the ensuing nine-page discussion Newton offers numerous objections to the wave theory (or rather wave theories). For example, in explaining various observed optical phenomena (such as refraction), wave theories suppose that these phenomena arise from the modification of light rays. (To explain the different degrees of refraction produced by a prism, wave theories claim that such differences are produced by the prism modifying the homogeneous rays of light, rather than by the prism separating heterogeneous rays). But Newton believes he has refuted "modification" theories by a series of experiments reported in Book I. Among Newton's other objections to the wave theory, perhaps the most famous is that if light "consisted in Pression or Motion either in an instant or in time, it would bend into the Shadow" (p. 362), as with water waves and sound waves. But no such bending (diffraction) had been observed.

The wave theory, various forms of which were supported by Newton's critics Hooke and Huygens, was the rival to Newton's corpuscular theory. That there are objections to such theories seems to be at least part of Newton's reasons for favoring a particle theory. Putting Queries 28 and 29 together, then, we might say that Newton is proposing a weak inference to a particle theory from two sets of considerations: the explanatory success of the particle theory (Query 29) and the objections to wave theories (Query 28). But how exactly is this inference supposed to proceed? What form does it take?

If Newton is making such an inference—and I think it is plausible to suppose that he is—he does not spell it out. On the basis of the discussions in Queries 28 and 29 themselves it is, I think, reasonable to say that Newton is arguing in some such way as this:

The hypothesis that light is corpuscular explains a range of observed optical phenomena.

The rival wave hypothesis has such-and-such difficulties.

Therefore (probably) light is corpuscular.

The inference is intended to be "weak," not "strong." But even so, *exactly* how it is supposed to go and whether it is reasonable even as a "weak" inference is not at all clear.

In what follows I will construct an idealized version of this argu-

ment which, although I cannot claim it to be Newton's, may nevertheless reflect some features of his thought.[51] The argument will have
two essential components: objections to the wave theory and an appeal to the explanatory power of the particle theory. Moreover, it will
be an argument whose conclusion is drawn with probability. This
probability will be reasonably high but not high enough to achieve
the certainty, or virtual certainty, of a "deduction from phenomena."
In constructing the argument I will assume that the usual axioms of
the probability calculus are satisfied. Probability can be construed
here as representing rational degress of belief.

Let us assume to begin with that light is either a particle phenomenon or a wave phenomenon. Newton himself offers no explicit argument for such an assumption, although here is one that he was in a
position to offer (and was in fact offered by his wave-theoretical
opponents in the nineteenth century).[52]

> Light is observed to travel in straight lines with uniform speed.
>
> In other cases, when something travels with uniform speed in a
> straight line this motion is always observed to be caused by a
> series of bodies or by a series of wave pulses produced in
> a medium (for example sound waves, water waves).
>
> Therefore, light is either a particle or a wave phenomenon.

Assume that the premises are true, that they report "phenomena,"
and that this is a "deduction from phenomena," so that the conclusion has the highest certainty possible in experimental philosophy.
Such an inference, a type of causal simplification, would be permitted
by Newton's Rule 2 requiring that like causes be assigned to like
effects, as far as possible. If in other cases motion is caused by particles or waves, then, unless there is evidence to the contrary, we
should infer like causes in the case of the motion of light.

Let T_1 be the hypothesis that light consists of particles, T_2 the hypothesis that light consists of waves, O the observed fact that light
travels in straight lines with uniform speed, and b the accepted background information which includes the information in the second
premise above. We might express the results of this argument
probabilistically as follows:

(1) $p(T_1 \text{ or } T_2/O\&b) = 1$.

That is, the probability that either T_1 or T_2 is true, given O and b,
is equal to 1.

Let us suppose that by appeal to certain other observed facts about
light—call them O'—we can show that the probability of the wave

theory is low, say less than one half. Which facts these are, and how this is to be shown, will be taken up in a moment. For the present let us simply write

(2) $p(T_2/O\&O'\&b) < 1/2$.

Now, if $p(T_1 \text{ or } T_2/O\&b) = 1$, then $p(T_1 \text{ or } T_2/O\&O'\&b) = 1$. Therefore from (1) and (2), since T_1 and T_2 are incompatible, we can infer

(3) $p(T_1/O\&O'\&b) > 1/2$.

Enter explanation. Let $O_1, \ldots, O_n$ be various observed facts about light (for example, reflection, refraction, variety of colors, fits of easy reflection and transmission) other than those in O and O' that are explained by the particle theory. And let us suppose that the explanations can be constructed in such a way that $O_1, \ldots, O_n$ follow deductively (in the ordinary sense) from the theory T_1, together possibly with the background information b. But if the particle theory T_1 (together possibly with b) deductively entails $O_1, \ldots, O_n$, then it follows from the probability calculus that $p(T_1/O_1, \ldots, O_n \&O\&O'\&b) \geq p(T_1/O\&O'\&b)$. So from (3), given the existence of such explanations, we derive

(4) $p(T_1/O_1, \ldots, O_n\&O\&O'\&b) > 1/2$,

which can be construed as the conclusion of the argument. It says that the particle theory is probable (more probable than not), given a range of observed optical phenomena, including ones explained by that theory.

The explanations of $O_1, \ldots, O_n$ provided by the particle theory do not create the high probability for that theory, but they do sustain it. They permit an inference from (3) to (4). This is an essential role played by such explanations in the attempt to establish high probability for the particle theory on the basis of a range of optical phenomena. To create high probability in the first place, a type of eliminative argument is used in which the wave and particle theories exhaust the probability, but the probability of the wave theory is low. How is the latter to be established?

Newton offers arguments of two types, one direct, the other indirect. An argument appealing to diffraction is an example of the former. In the case of other wave motions such as sound waves and water waves, diffraction into the shadow of an obstacle is observed, but no such diffraction into the shadow had been observed by Newton or others in the case of light (although Newton had observed diffraction away from the shadow).[53] So if we include in the background information the observed diffraction with other wave phenomena and the absence of such observations in the case of light, then the probability of the wave theory, given b, is low.

Second, Newton offers a more indirect type of argument. He points out that in order to explain certain observed optical phenomena, the wave theory introduces auxiliary assumptions that are either refuted by, or made very improbable by, observations. For example, to explain differences in refrangibility of rays emerging from a prism, wave theories introduce the auxiliary hypothesis that the prism modifies rather than separates the rays. Newton argues that this modification assumption is refuted or at least made extremely unlikely by further refraction experiments. (See Experiment 5, Book I, pp. 34ff.) Can we infer from this that the wave theory is improbable? We can if we suppose that the probability of the modification assumption, *given the wave theory* and the observations, is close to 1. Letting M be the modification assumption, T_2 the wave theory, and O' the results of various of Newton's refraction experiments, if $p(M/O\&O'\&b)$ is close to zero, and $p(M/T_2\&O\&O'\&b)$ is close to 1, then $p(T_2/O\&O'\&b)$ is close to zero. Even more generally, we get the same result if we suppose simply that $p(M/T_2\&O\&O'\&b)$ is much, much larger than $p(M/O\&O'\&b)$, without needing to suppose that the former probability is close to 1.[54]

Accordingly, there are two sorts of arguments to show that the probability of the wave theory is low—that is, (2). Once this is shown, we can infer the high probability of the particle theory (3) and its continued high probability in light of the various observed facts that it explains—that is, (4).

How "Newtonian" is the previous argument? In one respect quite un-Newtonian, since it explicitly invokes numerical probabilities and the probability calculus, neither of which, of course, Newton does. In certain other respects, however, it reflects what Newton seems to be doing in Queries 28 and 29. Assuming, as I have been, that Newton intends to provide some reasons for believing the particle theory, albeit "weak" ones, it gives the basis for inferring that theory with probability rather than certainty. Moreover, in doing so it takes into account Newton's criticisms of the wave theory and the explanatory virtues of the particle theory (each of which Newton himself emphasizes), showing how both contribute to the probability of the particle theory.

One objection that might be offered is that the preceding argument is a type of *eliminative* one, whereas Newton at one point rejects (a certain form of) eliminative reasoning. In a letter to Oldenburg of July 1672 he writes:

I cannot think it effectual for determining truth to examine the several ways by which phenomena may be explained, unless there can be a perfect enumeration of those ways. You know, the proper method for inquiring after the properties of things is to deduce them from experiments. And I told you that the theory, which I propounded [the theory of the heterogeneity of light rays] was evinced to me, not by inferring 'tis thus because not otherwise, that is, not by deducing it only from a confutation of contrary suppositions, but by deriving it from experiments concluding positively and directly.[55]

Newton here seems to be rejecting eliminative arguments of this form:

E: Each of the hypotheses $h_1, \ldots, h_n$, if true, will correctly explain phenomenon p.
But hypotheses $h_2, \ldots, h_n$ are false.
Therefore hypothesis h_1 is true.

Such arguments, which infer the truth of a hypothesis from the falsity of competitors, are fallacious, unless a complete enumeration can be made of all the competitors. Newton appears to be thinking of deductive interpretations of E in which if the premises are true, the conclusion must also be true. And he is correct in saying that arguments of form E, thus construed, are fallacious unless a complete enumeration of hypotheses is given. They are also fallacious if construed nondeductively—that is, as being such that the premises make the conclusion probable without entailing it. Unless some suitable assumption is made about the probability of the disjunction of hypotheses mentioned in the first premise, the conclusion that the probability of h_1 is high cannot be drawn.

    The particular eliminative argument I have constructed, however, is not of form E. The first step in the argument, which leads to (1), is not an explanatory step but one involving causal simplification. The claim is not that the particle and wave theories will both explain the finite rectilinear motion of light, but that in other observed cases when something travels with uniform speed in a straight line this motion is caused by a series of bodies or a series of wave pulses in a medium. Also, the claim in the first step is indeed exhaustive, since it assigns a probability of 1 to the disjunction of hypotheses. But even if it were not exhaustive in this sense, even if (1) were changed to read "$p(T_1$ or $T_2/O\&b)$ is close to but not equal to 1," a fallacy would not necessarily emerge (although other changes would need to be made in the argument).

I conclude that reconstructing what Newton does in Queries 28 and 29 in the form of a probabilistic argument that takes us from (1) to (4) is in conformity with certain important aspects of Newton's methodology. It combines his explanatory reasoning in Query 29 with his criticisms of the wave theory in Query 28 to provide some reason, though not the highest possible in experimental philosophy, to believe the corpuscular hypothesis. Although it is an eliminative argument, it is not one of the type Newton rejects. Because the hypothesis in question is inferred with a probability not sufficiently high to be a virtual certainty, Newton could not construe the argument as a "deduction from phenomena." While we should search for phenomena that will sanction such a "deduction," we should acknowledge that, assuming its premises to be true, the present argument does provide a legitimate "weak" reason for believing the corpuscular hypothesis.

vi. Strong versus Weak Inferences: An Assessment of One Tenet of Newtonian Methodology
A "strong" inference furnishes the proposition inferred with the highest evidence possible in experimental philosophy, a "weak" one with some evidence but not the highest possible. I shall suppose that this difference can be interpreted as a difference over probabilities (construed as representing degrees of rational belief). In a strong inference from A to B, the probability of B given A is close to, or equal to, 1. In a weak inference that probability is high (say greater than one half), but is not close to, or equal to, 1. For Newton both "strong" and "weak" inferences are based on "phenomena."

Now, I take it to be a tenet of Newtonian methodology that in experimental philosophy "deductions from phenomena," and only these, are strong inferences. Accordingly, there are two questions of assessment I want to raise: (1) Are Newton's "deductions from phenomena" *guaranteed* to be strong inferences? (2) Must other kinds of inferences fail to be strong?

To answer (1) we must return to the definition of "deduction from phenomena" offered in Section ii. Deductions from phenomena, we recall, include ordinary deductions, inductions, and causal simplification. Inductions are inferences from all observed members of a class to some members of the class that have not been observed, or to all members of the class. Let me simplify the discussion by considering deductive and inductive inferences but omitting causal simplification, which does not so readily lend itself to a general probabilistic treatment. Also, I shall discuss cases involving only deductions (in the ordinary sense) and those involving only inductions.

Deductive cases. Let $O_1, \ldots, O_n$ be descriptions of phenomena that together with background information b deductively imply h. Then $p(h/O_1, \ldots, O_n\&b) = 1$. So here we have a "strong" inference from the O's and b to h.

Inductive cases. To discuss these I shall first introduce some probability considerations, and afterwards apply them to the sorts of cases particularly relevant to Newtonian induction. Let h be a proposition that, together with background information b, deductively entails some observational statements $O_1, O_2, \ldots$ The following claims are provable.[56]

(a) If $p(h/b) \neq 0$, then $\lim_{n \to \infty} p(O_{n+1}/O_1, \ldots, O_n\&b) = 1$.

(b) If $p(h/b) \neq 0$, then $\lim_{m,n \to \infty} p(O_{n+1}\& \ldots \&O_{n+m}/O_1, \ldots, O_n\&b) = 1$.

Statement (a) tells us that if the prior probability of h is not zero, then as the number n of observed consequences of h and b gets larger and larger, the probability that the $n+1$ observational consequence of h and b will be true gets higher and higher, approaching 1 as a limit. Statement (b) tells us that if the prior probability of h is not zero, then as the numbers m and n get larger, the probability that the next m observational consequences are true, given that n observational consequences obtain, gets higher and higher, and approaches 1 as a limit.

To introduce the third probability result some restrictions will need to be made on h and O_i. Let h be some universal generalization of the form $(x)(Fx \supset Gx)$. Let the O's be "instances" of h of the form $Fa_i \supset Ga_i$. The following is provable:

(c) If (i) $p((x)(Fx \supset Gx)/b) \neq 0$, and (ii) $\lim_{n \to \infty} p((Fa_1 \supset Ga_1) \ldots (Fa_n \supset Ga_n)/b) = p((x)(Fx \supset Gx)/b)$, then $\lim_{n \to \infty} p((x)(Fx \supset Gx)/(Fa_1 \supset Ga_1) \ldots (Fa_n \supset Ga_n)\&b) = 1$.[57]

Statement (c) gives a set of sufficient conditions for the probability of $(x)(Fx \supset Gx)$, given observed instances of the form $Fa_i \supset Ga_i$, to approach 1 as a limit.

Now let us apply these three probability results to Newtonian inductions. In all three cases let us consider h's of the form $(x)(Fx \supset Gx)$, and O's instances of the form $Fa_i \supset Ga_i$. Statement (a) tells us that if the prior probability of $(x)(Fx \supset Gx)$ is not zero, then as the number of observed instances of $(x)(Fx \supset Gx)$ increases, the probability that the next instance will obtain gets higher and higher, approaching 1 as a limit. A similar claim can be made for (b). Statements (a) and (b), so construed, correspond to Newton's inductions from some observed members of a class to some other member(s) of that class. Statement (c) corresponds to Newton's inductions from some observed members of a class to all members. In all three cases the probability in question approaches 1 as a limit, under certain very weak assumptions. Intuitively, the probability that the next instance will satisfy $(x)(Fx \supset Gx)$, that the next m instances will, and that all instances will gets higher and higher as more and more instances are observed. We get more and more certainty in these cases with more and more observed instances of $(x)(Fx \supset Gx)$.

Of course, it is not the case that for every number n of observed instances the probability that the next instance will satisfy $(x)(Fx \supset Gx)$, that the next m instances will (for any m), and that all instances will is close to 1. Consider just the latter, and suppose that the prior probability of $(x)(Fx \supset Gx)$ is low, and the O's are such that, with sufficiently small n, the prior probability of the conjunction of O's is high. If $(x)(Fx \supset Gx)$ and b entails the O's, then by Bayes's theorem,

$$p((x)(Fx \supset Gx)/O_1, \ldots, O_n \& b) = p((x)(Fx \supset Gx)/b)/p(O_1, \ldots, O_n/b).$$

Now if $p((x)(Fx \supset Gx)/b)$ is low and the O's are such that, with sufficiently small n, $p(O_1, \ldots, O_n/b)$ is high, then the probability on the left will be small, despite the fact that all the observed O's satisfy $(x)(Fx \supset Gx)$. One type of case of this sort involves Goodmanesque properties such as "grue," where the prior probability of the proposition "All emeralds are grue" is very low, but where, given appropriate background information, the probability that observed items are grue if emeralds is very high. Strange Goodmanesque properties or classes can prevent the probability on the left from being high for a given n. But as n increases without bound, the probability on the left will approach 1 as a limit, strange properties notwithstanding.

Goodmanesque properties are not the only things that can prevent the probability on the left from being high for a given n. Recall the proof of Proposition I of the *Opticks*, in which an induction is made from observations of differences in refrangibility of blue and red rays in an experiment with the sorts of prisms used by Newton to differences in refrangibility of any differently colored rays in any sort of refraction, whether or not the latter is produced by a prism. A critic of Newton might argue as follows: (i) the number n of observed instances of Proposition I (that lights differing in color differ in degrees of refrangibility) is quite low. (If we count as instances here the results of *types* of experiments, rather than specific trials, then the critic has some justification, since Newton cites only two experiments.) (ii) The critic might agree that the probability of getting the observed results Newton obtains *with these types of experiments with prisms* is high, while supposing that analogous refraction results with other sorts of prisms or without prisms are improbable.[58] (iii) The critic might argue, on the basis of background information b he possesses, that the prior probability of Newton's Proposition I is very low. At least, the critic might argue, Newton does nothing to dispel doubts expressed in (i)–(iii). But unless such doubts are removed, the probability of Newton's Proposition I, given the results of the experiments Newton mentions, cannot be assumed to be high. The most we can say is that this probability will increase toward 1 as the number of observed instances of the proposition increases.

Confining our attention to ordinary deductions and inductions, we can now answer the question "Are Newton's 'deductions from phenomena' guaranteed to be strong inferences?" in the following way. If they are deductions (in the ordinary sense), they are so guaranteed. Any deductive inference from $O_1, \ldots, O_n$ and b to h, no matter what number n is, guarantees that the probability of h, given $O_1, \ldots, O_n$ and b, is maximal. By contrast, it is not the case that every particular inductive inference is guaranteed to be strong, no matter how many instances are involved and no matter what the character of the properties or classes in question. If Newton's methodology requires a claim to the contrary, then it is mistaken. The previous probability results show, however, that (under certain weak assumptions) as more and more instances are observed, then no matter what the character of properties or classes in question, the strength of the inference is guaranteed to increase and to approach the highest strength in the limit.

Accordingly, there are several ways to interpret Newton's methodology (or to modify that methodology) so as to avoid the problems just mentioned. First, instead of saying that every inductive inference from phenomena is a strong one, Newton could say that *some* are, specifically those based on sufficiently many instances—provided that the prior probability of $(x)(Fx \supset Gx)$ is not zero. Second, Newton could restrict those inductions he will allow in the category of "deductions from phenomena" to ones that are based on sufficiently many instances. On both these proposals, however, no particular number can be given in general that will count as "sufficiently many." In each case this will depend on the prior probability of $(x)(Fx \supset Gx)$ and on the prior probability of the conjunction of instances. Third, Newton could attempt to impose conditions on the character of the properties or classes that are subject to induction in such a way that inductions involving such properties or classes will guarantee maximal probability no matter how many instances have been observed. Newton does not formulate any such conditions. Whether it would be possible to do so seems very dubious to me, though I shall not pursue this here. Finally, Newton could abandon entirely an "absolute" claim about the strength of inductions in favor of a "comparative" one. He could say simply that, under minimal assumptions, the strength of an induction increases as more and more instances are observed.

Now, turning to the other side of the coin, we need to ask whether in experimental philosophy there are strong inferences that are not "deductions from phenomena." Is Newton correct in implying that only "deductions from phenomena" can have this feature?

Let us return to result (a) above. (What I say here will be applicable to (b) as well, *mutatis mutandis*.) Although (a) allows h and O_i to be of forms $(x)(Fx \supset Gx)$ and $Fa \supset Ga$ respectively, it does not require this. All that is necessary is that h and b deductively imply O_i. Accordingly, h might be some proposition that Newton would classify as a "hypothesis," for example, that light consists of particles (h). This hypothesis is not "deduced from phenomena." Let the background information b include Newton's first law of motion that in the absence of forces particles travel with uniform speeds in straight lines. Hypothesis h together with b deductively implies (O_1) that in the absence of forces light travels in straight lines, and (O_2) that in the absence of forces light travels with uniform speed. Now result (a) allows us to conclude that the probability that some consequence of a "hypothesis" (in the Newtonian sense) obtains gets higher and higher, approaching 1 as a limit, as more and more consequences of that hypothesis are observed. The only assumption needed is that the prior probability of this hypothesis is not zero. This, of course, does not imply that the probability of the "hypothesis" itself approaches certainty, but only that the probability of its deductive consequences does.

Newton does not appear to be thinking of cases in which we make inferences to deductive consequences of "hypotheses." But such inferences can be strong ones, or at least get stronger and stronger as more and more consequences are observed to hold. To be sure, Newton could claim that he is classifying as "inductive" an inference from some observed consequences of h to other not yet observed consequences. But his inductions appear to be simply inductive generalizations from observed F's that are G's to other or all F's being G's.

Let us turn, then, to result (c) involving the probability of h itself. And let us consider the more general case in which h is any proposition that, together with b, deductively implies $O_1, O_2, \ldots$ Here we cannot obtain the result that $\lim_{n \to \infty} p(h/O_1, \ldots, O_n \& b) = 1$ because we cannot in general assume that $\lim_{n \to \infty} p(O_1, \ldots, O_n/b) = p(h/b)$. Indeed, the following are provable:

(d)   Let h (together with b) entail O_1, O_2, . . . If h has at least one incompatible competitor h' that together with b also entails O_1, O_2, . . . , and whose probability on b is greater than zero, then $\lim_{n \to \infty} p(h/O_1, \ldots, O_n \& b) \neq 1$.

(e)   Let h together with b entail O_1, O_2, . . . If h has at least one incompatible competitor h' that together with b also entails O_1, O_2, . . . , and is such that $p(h'/b) \geq p(h/b)$, then for any n, no matter how large, $p(h/O_1, \ldots, O_n \& b) \leq 0.5$.[59]

So if h has competitors that, like h, deductively imply all the observable phenomena, then h's probability will not approach 1 as a limit; and if the prior probability of one of the competitors is at least as great as h's prior probability, then h's probability will not increase beyond 0.5, no matter how many deductive consequences of h are observed to be true.

The quest for strong inferences to Newtonian "hypotheses" is not necessarily doomed, however. We need not insist that the limit of the probability of h be 1, but only that the probability of h given the observations be "very high" and remain so with more and more observations. To this end I shall employ the concept of a *partition* of propositions on b, which is a set of mutually exclusive propositions the probability of whose disjunction on b is 1, and the probability of each of which on b is not zero. The following is provable.[60]

(f)   If h, h_1, . . . , h_k form a partition on b, then for any O, and for each h_i ($\neq h$) in the partition, and for any number r greater than or equal to zero and less than 1, $p(h/O \& b) > r$ if and only if $\Sigma_{i=1}^{k} p(h_i/ O \& b) < 1 - r$.

Now, suppose that we have some observed phenomena O, O_1, . . . , O_n and background information b, and we want to make a strong inference to h by showing that the probability of h, given the observed phenomena and background information, is greater than some threshold value r for "very high" probability. Using theorem (f), the following strategy is possible:

> *Strategy for showing that* h *has a very high probability (greater than some threshold value* r *for very high probability), given observed phenomena* O, O_1, . . . , O_n *and background information* b:

1. Find some partition on b—h, h_1, . . . , h_k—that includes h.
2. Show that phenomenon O is such that for each proposition $h_i \neq h$ in the partition, $\Sigma_{i=1}^{k} p(h_i/O \& b) < 1 - r$.
3. Show that O_1, . . . , O_n are derivable from h (together with b).

If we complete steps 1 and 2 in this strategy, then, in accordance with theorem (f), we will have shown that $p(h/O\&b) > r$. By completing step 3, we show that $p(h/O\&O_1, \ldots, O_n\&b) > r$, since $O_1, \ldots, O_n$ are derivable from h together with b.

The question of interest is whether this strategy is applicable to propositions Newton would regard as "hypotheses." In fact, it seems applicable to the hypothesis Newton considers in Query 29 of the *Opticks*: that light consists of particles. Indeed, the probabilistic argument I constructed in the last section can be suitably modified and shown to be a legitimate variant of this form. Let me recall the basic steps.

We began with the observation O that light travels in straight lines with uniform speed, which, together with the background information b, yields a probability of 1 that light is corpuscular or undulatory. That is,

(1) $p(T_1 \text{ or } T_2/O\&b) = 1$.

Accordingly, T_1 and T_2 form a partition on $O\&b$, since T_1 and T_2 are incompatible.

Second, we found some other observed facts O' that cast doubt on the wave theory T_2. We noted this by writing $p(T_2/O\&O'\&b) < 1/2$. But this is too modest, even for Newton, since the actual facts cited, Newton thought, cast much more doubt on T_2 than this. The two mentioned were diffraction and refraction. If light is a wave phenomenon, then, like water waves and sound waves, it ought to be diffracted into the shadow; but no such diffraction was observed by Newton. Also, Newton (as well as defenders of the wave theory) believed that, *given the wave theory* and the observations of differences in degrees of refraction, the probability that light is modified by the refracting prism is very high, say close to 1. But on the basis of his own refraction experiments Newton pretty clearly thought he had refuted the modification assumption; that is, the probability of this assumption, given his experimental results, is close to zero. So, where M is the modification assumption, and O' includes the results of Newton's refraction experiments, we have the result that $p(M/T_2\&O'\&O\&b)$ is close to 1, whereas $p(M/O'\&O\&b)$ is close to zero. It follows that $p(T_2/O'\&O\&b)$ is close to zero. Letting O' also contain the observed absence of diffraction into the shadow and b also contain observed diffraction in the case of sound and water, we write

(2) $p(T_2/O\&O'\&b) \approx 0$ ($\approx$ means "is close to").

This completes the second step in the strategy.

From (1), since the probability of T_1 or T_2 is 1, it remains 1 if we add O'. So we have

(3) $p(T_1 \text{ or } T_2/O\&O'\&b) = 1$.

Since T_1 and T_2 are incompatible, from (2) and (3) we infer

(4) $p(T_1/O\&O'\&b) \approx 1$.

Since T_1 and b deductively imply other optical phenomena $O_1, \ldots, O_n$, from (4) we derive

(5) $p(T_1/O\&O'\&O_1, \ldots, O_n\&b) \approx 1$,

which completes the third and final step of the strategy.

Again, I must stress that it is not my claim that this is Newton's actual argument in Queries 28 and 29. Besides the attribution of the probability calculus, the main stumbling block lies in the use of the first step in the strategy, leading to (1). Although Newton considers only the wave and particle theories, he does not explicitly claim that the probability of their disjunction on the evidence is maximal; he does not argue, in effect, that these hypotheses form a partition. Still, in the previous section I indicated what type of argument for this claim Netwon could have given that would be compatible with his general methodology. If the strategy is launched by this assumption in step (1), then Newton's own arguments against the wave theory can be used to justify the steps leading to the final (5).

Let us suppose that (1) is justified by inference from observed phenomena. And let us assume that the remaining steps are also valid, so that the argument does establish the very high probability of a proposition, given certain observations and background information. If so, it provides the basis for a "strong" inference to that proposition from those observations and background information. Is the argument a "deduction from phenomena"?

In certain ways it seems quite different from the sorts of arguments Newton has in mind when he uses this expression. First, unlike the "deductions" that Newton gives, it contains an inference to a disjunction of propositions in the first step. Second, the argument is eliminative, whereas the "deductions" Newton offers are not. Indeed, he rejects (certain types of) eliminative arguments. Third, and most important, it makes use of the probability calculus, which Newton never does. The inferences to (3), (4), and (5) are justified by principles of probability. Whether Newton would have been willing to classify such inferences as "deductive" is unclear.

Yet reasons might be offered for classifying the argument as a "deduction from phenomena." First, the previous characterization of causal simplification (as well as that of induction) does not preclude an inference to a disjunction of propositions. Second, although it is eliminative, it is not an eliminative argument of the type that Newton rejects. Indeed, if the previous point is accepted, it is an eliminative argument that uses causal simplification to establish a disjunction of propositions and then to argue against one of the disjuncts. Third, the probability principles generating steps (3), (4), and (5) might be thought of as, or as akin to, mathematical principles, which for Newton can serve as a basis for "deductions."

Accordingly, assuming the argument in question is valid, the following possibilities emerge:

1. In a broad sense the argument is a "deduction from phenomena." If so, it does not refute the Newtonian claim that only "deductions from phenomena" guarantee strong inferences. If we construe it as a "deduction from phenomena," however, then with this argument we must deny that the Newtonian corpuscular hypothesis is a hypothesis. With this argument we will have "deduced" the corpuscular hypothesis from the phenomena and thus rendered it no longer hypothetical.

2. In a narrower sense (one that excludes probability arguments) the argument is not a "deduction from phenomena." Yet it provides the basis for a "strong" inference to the corpuscular hypothesis. So if this narrower sense is Newton's, then we need to reject his idea that only "deductions from phenomena" can provide the highest certainty in experimental philosophy.

vii. Conclusions

1. Although neither Newton's professed methodology, nor his actual practice, forms a consistent set, my suggestion is that interpretation (C) in Section iii reflects a good deal of both. On that interpretation, the most certainty possible in experimental philosophy is achieved when, and only when, propositions are "deduced from phenomena." The latter involves deduction or induction or causal simplification from generally accepted facts established by observation.

2. On this interpretation, however, one is allowed not only to consider propositions not "deduced from phenomena" (that is, hypotheses) but also to make weak inferences to them in cases in which "deductions" have not been achieved. But we must recognize that such inferences are weak, and we must continue to search for phenomena from which the propositions in question can be "deduced."

3. One sort of non-"deductive" inference to hypotheses is illus-

trated in Queries 28 and 29 in the *Opticks*, where Newton discusses the particle and wave theories of light. Here he seems to be making a (weak) inference to the particle theory on the grounds that it explains a range of optical phenomena. In Section v this argument is reconstructed probabilistically in such a way as to reflect, at least in part, Newton's discussion in Queries 28 and 29, as well as his general methodology.

4. We cannot suppose, as Newton seems to, that "deductions from phenomena" will always yield the maximal certainty possible in experimental philosophy. In the case of induction, for example, such certainty is not guaranteed simply by observing positive instances of an inductive generalization and no negative ones. What we can say is that, granted certain minimal assumptions, an increase in the number of positive instances will increase the strength of such inferences toward maximality. Finally, assuming that probabilistic explanatory reasoning of the type constructed in Section vi can be valid, we may say this: If probabilistic arguments are not construed as "deductive," then we cannot suppose, as Newton seems to, that only "deductions from phenomena" can generate the highest certainty possible in experimental philosophy.[61]

Notes

1. *Opticks* (New York: Dover, 1979), p. 370.
2. *Principia* (University of California, 1966), p. 547.
3. *Newton's Philosophy of Nature*, ed. H. S. Thayer (New York, 1953), p. 6.
4. Ibid., p. 7.
5. *Opticks*, p. 7.
6. Ibid., p. 404.
7. From a manuscript sheet translated by J. E. McGuire, "Body and Void in Newton's De Mundi Systemate: Some New Sources," *Archive for History of Exact Sciences* 3 (1966), pp. 238–239. There are two other unpublished definitions of *phenomena* that are substantially the same as this. In addition to the quoted definition there is also an unpublished Rule 5 which contains a brief discussion of phenomena that conforms to the unpublished definition. See I. Bernard Cohen, *Introduction to Newton's Principia* (Cambridge, Mass., 1978), p. 30.
8. They are listed as such in the second and third editions, but not in the first, where they are called hypotheses. For a comment on Newton's terminological change, see n. 30.
9. *Principia*, p. 404.
10. In a letter to Cotes in 1713, written as the second edition of the *Principia* was being prepared for publication, Newton explicitly counts induction as a method of deduction (indeed, in this passage, as the only method): "experimental philosophy proceeds only upon phenomena and deduces general propositions from them only by induction." It is difficult to believe, however, that for Newton the only deductions from phenomena are inductions. Recall that at the end of the *Principia* he writes that "particular propositions are inferred from the phenomena and *after-*

wards rendered general by induction," which suggests that there are noninductive inferences to begin with and then inductive ones to generalize from these. Newton's remarks to Cotes in the previous passage pertain to "general" propositions. So perhaps he was thinking here only of those propositions (such as the laws of motion, which he mentions) that do require generalization by induction.

Descartes also uses "deduction" to include induction as well as deduction. See Desmond Clark, *Descartes' Philosophy of Science* (Penn State, 1982), chap. 3.

11. Ernan McMullin, *Newton on Matter and Activity* (Notre Dame, 1978); see pp. 13ff. Maurice Mandelbaum, *Philosophy, Science, and Sense Perception* (Baltimore, 1964); see pp. 74ff.

12. *Principia*, p. 399.

13. J. E. McGuire, "The Origins of Newton's Doctrine of Essential Qualities," *Centaurus* 12 (1968), pp. 233–260; see p. 244, pp. 245–246. McGuire argues that Newton held that qualities not subject to "intension and remission" are *essential* qualities of bodies.

14. See McGuire, "Origins," p. 237.

15. McMullin, op. cit., pp. 11–12.

16. *Ibid.*, pp. 12–13.

17. Indeed, in his first version of the rule Newton omits the "intension and remission" clause and writes simply: "The laws (and properties) of all bodies on which it is possible to institute experiments, are laws (and properties) of all bodies whatsoever." See McGuire, "Origins," p. 236.

18. A somewhat different interpretation of Newton might be given as follows. By "induction" Newton means an inference from observed members of a class to unobserved but observable members. Where induction is concerned, there is no restriction to properties that do not admit of "intension and remission." Rule 3, however, is not a (straightforward) inductive rule but a more powerful one. It sanctions inferences from observed members of a class to all members of that class, including the unobservable ones. And when one makes an inference from the observed to the unobservable, a restriction is required to properties that cannot be "intended and remitted." Such an interpretation seems to be the one, or at least close to the one, offered by McGuire in "Atoms and the 'Analogy of Nature': Newton's Third Rule of Philosophizing," *Studies in History and Philosophy of Science* 1 (1970), pp. 3–58; see p. 12. It is in conflict with interpretations of Mandelbaum (op. cit., p. 62) and McMullin (op. cit., p. 15), both of whom construe Rule 3 as an inductive rule which permits inferences from the observed to the unobservable. I take the latter position in what follows. Newton does not explicitly say that inductions are restricted to the observable. And, as I noted earlier, he does use the term *induction* in referring to the sorts of inferences he illustrates in Rule 3. In any event, on both the interpretation suggested by McGuire and the one I shall adopt, Newton is not to be construed as restricting inductions to cases involving properties that cannot be "intended and remitted."

19. Rule 4 states: In experimental philosophy we are to look upon propositions inferred by general induction from phenomena as accurately or very nearly true, notwithstanding any contrary hypotheses that may be imagined, until such time as other phenomena occur by which they may either be made more accurate or liable to exceptions.

20. Reprinted in J. E. McGuire, "Newton's 'Principles of Philosophy': An Intended Preface for the 1704 *Opticks* and a Related Draft Fragment," *British Journal for the History of Science* 18 (1970), pp. 178–186.

21. William Harper emphasizes the idea of unification here rather than simplicity. He

thinks of Newton's first two rules as endorsing a policy of unifying natural kind conceptions wherever possible. See his "Newton's Unification of Heaven and Earth" (draft ms.).

22. McGuire, "Atoms," thinks that after 1690 Newton regarded only his laws of motion and the eight axioms in Book I of the *Opticks* as deduced from phenomena. This strikes me as too strong. In that same 1713 letter to Cotes, Newton does cite the laws of motion as examples of propositions deduced from phenomena by induction; but he also cites the law of gravitation and the proposition that bodies are impenetrable, mobile, and exerters of force.

23. *Principia*, p. 399.

24. McGuire, "Atoms," calls inferences that proceed from "what is observable to what is in principle unobservable" *transductions* (p. 3), which (by contrast to Mandelbaum and McMullin), he wants to distinguish from inductions. My claim is that Rule 3 allows Newton to make inferences to properties of all members of a class, even if some members and their properties are unobservable. By itself it does not allow Newton to infer that unobservable members of the class exist (just because observable ones do).

25. *Principia*, p. 399.

26. In a 1713 letter to Cotes, Newton does use the expression "experimental proof" in such a way as to suggest that it has the same meaning as "deduction from phenomena." See the second quote at the beginning of this paper.

27. *Opticks*, p. 21.

28. In *The Newtonian Revolution* (Cambridge, 1980), I. B. Cohen claims that Newtonian methodology in the *Opticks* is significantly different from that in the *Principia* (see pp. 13ff.). Cohen points out that although Newton included axioms and definitions in both, he makes use of these only in the *Principia*, not in the *Opticks*, where the proof is "by experiments." Moreover, Cohen claims that Newton's methodology in the *Principia* (which Cohen calls the "Newtonian style") involves a process of mathematical idealization and simplification foreign to the *Opticks*. This process begins with a set of idealized physical entities and conditions that can be expressed mathematically; deductions are made and compared with data and observations. This leads to a modification of the original assumptions to form a more complex mathematical system and to further deductions and comparisons with nature (see pp. 62–68). "In the *Principia*," writes Cohen, "the role of induction is minimal" (p. 16). I have three comments on this.

1. Cohen is certainly correct when he points out that in the *Opticks*, when Newton "proves" his propositions, he does not make explicit reference to the axioms and definitions, as he does in the *Principia*. The work is considerably less formal and precise. But the proofs work—that is, they demonstrate what they are supposed to—only if the definitions and axioms are implicitly assumed. For example, to infer Proposition I in the *Opticks* from the facts reported in Experiment 1, Newton is making at least implicit use of his definition of *refrangibility* as the disposition of rays to be refracted or turned out of their way in passing out of one transparent body or medium into another (p. 2). And unless he relates refraction to how a refracted body appears—as he does in Axiom VIII and in the discussion that follows—in his proof of Proposition I he will not be able to infer anything about differences in degrees of refrangibility from differences in the displacements of the observed positions of the red and blue colors on the paper.

2. Newton himself regarded his axioms of motion in the *Principia* as "deduced from the phenomena and made general by induction." That is, he regarded induction as crucial for generating his three most basic laws. Moreover, he makes explicit

reference to induction in the proofs of various propositions in Book III. As far as mathematical idealizations go, it is true that many of the propositions in the *Principia* are concerned with systems that do not exactly correspond to any known in nature. For example, Proposition I in Book I deals with one body subject to a single central force. But Newton derives this proposition by deduction from his laws of motion plus their corollaries. He does not begin in this case with an *actual* system such as a planet and the sun which he then idealizes by ignoring forces on the planet exerted by other planets. The question is simply: What would happen if a body were subject to a central force? The answer is supplied by deduction from other propositions.

    3. Finally, as McGuire has emphasized, if we examine Newton's intended preface for the *Opticks*, we will conclude that Newton himself "did not see any dichotomy in method between that used in the *Principia* and that found in the *Opticks*." McGuire, "Intended Preface," p. 182.

29. For example, in his discussion of Proposition VI of Book I, Newton offers a demonstration of the sine law that involves introducing what he himself call a "Supposition"—that bodies refract light by acting upon its rays in lines perpendicular to their surfaces. Newton makes clear in what follows that he means that a *force* is exerted by the refracting medium on the rays of light. Now, for Newton a force can only act upon a *body*. (In the *Principia* Newton's definition of an impressed force begins with the phrase "an action exerted *upon a body*.") So it seems that here Newton is treating the rays of light as bodies. Yet the claim that the rays are bodies (which strongly suggests the corpuscular theory) is not something that Newton "deduces from the phenomena." Despite this, Newton does not seem to regard the "Supposition" as a mere supposition, since he takes the resulting demonstration "to be a very convincing Argument of the full truth" of the sine law (p. 82).

30. See I. B. Cohen, *Franklin and Newton* (Philadelphia, 1956), pp. 138–140; Cohen, "Hypotheses in Newton's Philosophy," *Physis* 8 (1966), pp. 163–184; Alexander Koyré, *Newtonian Studies* (Chicago, 1965), pp. 261–272; N. R. Hanson, "Hypotheses Fingo," in Robert E. Butts and John W. Davis, eds., *The Methodological Heritage of Newton* (Toronto, 1970), pp. 14–33. We need only be reminded that what in the second and third editions of the *Principia* are called phenomena are called hypotheses in the first edition. According to Koyré, op. cit., p. 31, in the first edition by contrast to later ones Newton means by "hypothesis" any fundamental assumption of the theory (so that the laws of motion would be hypotheses, in this sense, even though, according to Newton, they are "deduced from phenomena").

31. *Principia*, p. 547.

32. *Newton's Philosophy of Nature*, p. 6.

33. *Principia*, p. 547.

34. See Hanson, op. cit., p. 14.

35. *Newton's Philosophy of Nature*, p. 6.

36. Here I disagree with Koyré (op. cit., pp. 36–37), who says that "hypothesis" (in the sense under discussion) "means for Newton something that cannot be proved." Whether Koyré means this in sense (b) or (c) is not clear.

37. *Principia*, p. 547.

38. In a letter to Boyle of February 1678 (*Opera*, IV, pp. 385–394), Newton speculates a great deal more about this subtle spirit, or aether, even offering a hypothesis about the cause of gravity: "I shall set down one conjecture more, which came into my mind now as I was writing this letter: it is about the cause of gravity. For this end I will suppose aether to consist of parts differing from one another in subtlety by indefinite degrees" (p. 394). Newton indicates considerable hesitation in proposing

hypotheses to Boyle, and he is clearly not saying that such hypotheses are established. Yet equally clearly he is not avoiding them.

39. *Principia*, appendix, p. 673.

40. *Principia*, appendix, p. 673.

41. *Principia*, p. 547.

42. See Robert Hugh Kargon, *Atomism in England from Harriot to Newton* (Oxford, 1966), pp. 107, 108, 124.

43. Even if (C) is more adequate than other interpretations, it must not be assumed that it best reflects Newton's overall views and practice *throughout his professional career*. It is quite possible that by the later 1690s in response to criticisms of the first edition of the *Principia* not only did Newton's use of the term *hypothesis* change but also his ideas about hypotheses (in the sense in question) became more sharply articulated. See Cohen, "Hypotheses in Newton's Philosophy," p. 179.

44. *Opera*, IV, p. 324.

45. *Opera*, IV, p. 394.

46. *Opticks*, pp. 404–405.

47. In the nineteenth century Henry Brougham, a follower of Newton in both professed methodology and in his defense of the particle theory of light, also speaks of analysis and synthesis. Brougham insists that the phenomena explained in synthesis must be such as to provide a sufficient basis for an inductive argument to the explanatory proposition. Perhaps this is what is meant, or part of what is meant, by "proving the explanation." In the passage quoted, however, Newton does not explicitly require what Brougham does.

48. *Opticks*, p. 405.

49. Cohen, *Franklin and Newton*, p. 184, so interprets Newton.

50. A. I. Sabra, *Theories of Light from Descartes to Newton* (London, 1967), p. 312 (emphasis in original).

51. See my "Light Hypotheses," *Studies in History and Philosophy of Science* 18 (1987), pp. 293–337.

52. See ibid.

53. See Roger H. Stuewer, "A Critical Analysis of Newton's Work on Diffraction," *Isis* 61 (1970), pp. 188–205.

54. See "Light Hypotheses."

55. *Newton's Papers and Letters*, ed. I. B. Cohen, p. 93.

56. See John Earman, "Concepts of Projectibility and Problems of Induction," *Nous* 19 (1985), pp. 521–535.

57. For proof, see ibid., p. 529.

58. See Simon Schaffer, "Glass Works: Newton's Prisms and the Uses of Experiment," in David Gooding, Trevor Pinch, and Simon Schaffer, eds., *The Uses of Experiment* (Cambridge, 1989).

59. See Earman, "Concepts," pp. 528–529.

60. See my "Hypotheses, Probability, and Waves," forthcoming in *British Journal for the Philosophy of Science*.

61. Research for this paper was supported by a grant from the NEH. For very helpful suggestions I am indebted to Robert Rynasiewicz, Doren Recker, Robert Kargon, and Alan Shapiro.

Chapter 9

Reason and Experiment in Newton's *Opticks:* Comments on Peter Achinstein

R. I. G. Hughes

Peter Achinstein gives us a very careful account of key terms in Newton's methodological writings, terms such as *phenomena, deduction,* and *hypothesis,* together with an ingenious reconstruction of a "Newtonian" argument by induction in favor of the corpuscular theory of light.

I will address myself to his paper rather obliquely, by presenting an alternative view of Newton's "experimental philosophy," pointing out as I go along where Achinstein and I differ, at least in emphasis. I will limit myself to the way experimental philosophy appears in Newton's *Opticks*,[1] and leave aside the question whether this is consistent with what goes on in the *Principia*.[2]

In the *Opticks* we find Newton's methodology exemplified in the early books, and described in the last pages of the Queries; similar views are set out in a draft fragment of about 1703, which McGuire suggests was written for possible inclusion in the *Opticks*.[3] I will look first at *induction*, which, as Newton tells us, plays a key role in scientific method. In Query 31 he says that scientific investigation starts by "making experiments and observations and drawing general conclusions by induction" (p. 404). This affords "no demonstration," he goes on—that is to say, it does not have the certainty of what we would call mathematical proof—but it is "the best way of arguing which the nature of things admits of."

But just how good does Newton take this "best way of arguing" to be? Achinstein suggests that it can establish propositions "beyond reasonable doubt," and his suggestion finds support from Newton's fourth rule of reasoning in philosophy in the *Principia* (p. 400): "In experimental philosophy we are to look upon propositions inferred by general induction from phenomena as accurately or very nearly true."[4] Newton's actual practice in the *Opticks,* however, is much less confident, as we can see from his treatment of the phenomenon known as "Newton's rings" in Part 1 of Book II.

When a lens is placed on a mirror and viewed from above, a series of rings of different colors appears, centered on the center of the lens.

The diameter of a given ring changes with the observer's angle of viewing: "The rings were least, when my eye was placed perpendicularly over the glasses in the axis of the rings: and when I viewed them obliquely they became bigger, continually swelling as I removed my eye farther from the axis" (pp. 203–204). Now, except where the two touch each other, between the lens and the mirror there is an air gap whose thickness increases with its distance from their point of contact; by measuring the diameter of a ring Newton can calculate the thickness of the air gap where it is produced. In this way he investigates how the thickness of the gap for a given ring varies with the angle of viewing. He concludes: "And from these measures I seem to gather this general rule: that the thickness of the air is proportional to the secant of an angle whose sine is a certain mean proportional between the sines of incidence and refraction" (p. 205).

I take this to be an example of a general conclusion reached by induction; in fact Newton tabulates readings at thirteen different angles. But note the tentative way it is expressed: "From these measures I seem to gather this rule." This is not the tone of voice of one who thinks the conclusion has been established beyond all reasonable doubt. Within half a dozen pages there are three more conclusions, each expressed with a similar diffidence; we find, on p. 207, "Perhaps it may be a general rule that . . . ,"[5] and on pp. 210 and 211, "Thence it may be gathered that . . . ," and, "And hence I seem to collect that . . ."

There are good reasons for this diffidence. The last of these quotations continues: "I seem to collect that the thicknesses of the air [at the] limits of the five principal colors (red, yellow, green, blue, violet) . . . are to one another very nearly as the sixth lengths of a chord which found the notes in a sixth major, *sol, la, mi, fa, sol, la*" (pp. 211–212). Newton is here exploring a possible connection between the colors of the spectrum and the notes in a diatonic scale. But then he writes: "But it agrees something better with the observations to say, that the thicknesses [at the] limits of the seven colors, red, orange, yellow, green, blue, indigo, violet in order, are to one another as the cube roots of the squares of the eight lengths of a chord, which found the notes in an eighth, *sol, la, fa, sol, la, mi, fa, sol*" (p. 212).

In other words we can have *competing* inductions: there may be more than one general rule under which we can subsume the phenomena. In the case at hand, which of the rules—call them postulates A and B—should we choose? This is the problem of projectibility, Goodman's problem as opposed to Hume's. It can appear in (at least) two different forms. The first is that, given any finite number of points illustrating the dependence of one variable on another, there is an infinite number of analytic functions that they may exemplify. The second is that in practice we deal not with points but with splotches. It is the second that is operative here; Newton says, "It agrees something better with the observations" to advance postulate B, but this is not to say that, given sufficient theoretical pressure, we might not want to remain with postulate A. Where, within Newton's experimental philosophy, could such theoretical pressure come from?

Query 31 of the *Opticks* does not just tell us that in natural philosophy we need to make experiments and to draw conclusions by induction; it also tells us that this is only half of what is involved in scientific theorizing, the half known as *analysis*. And, as we have just seen, the moral to be drawn from Book II of the *Opticks* is that such inductions, taken on their own, are not conclusive. The other half of science is *synthesis*, or *composition* (p. 403); the propositions of science must be set out not as an unordered collection of inductive generalizations but in a way that establishes the logical connections between them. To use an anachronism, only when we do this do we have a *theory*.

Newton was aware, incidentally, that this two-stage model of theory construction was oversimplified; in a draft fragment of about 1703 he describes a more intricately dialectical process:

> The method of resolution [analysis] consists in trying experiments and considering all the phenomena of nature relating to the subject in hand, and examining the truth of those conclusions by new experiments and drawing new conclusions from those experiments, and so proceeding alternately from experiments to conclusions and from conclusions to experiments until you come to the general properties of things (and by experiments and phenomena have established the truth of those properties). Then assuming those properties as principles of philosophy you may by them explain the causes of such phenomena as follow from them; which is the method of composition.[6]

It is not known why this passage did not find its way into the *Opticks*; the oversimplifications inherent in the two-stage model, however, do not affect the point I am making here, since on both accounts composition is portrayed in the same way, as the deduction (in our sense) of a large number of propositions from a few basic principles.

With this specification in mind, let us look at Book I of the *Opticks*. It begins with the words: "My design in this book is not to explain the properties of light by hypotheses, but to propose and prove them by reason and experiments." Reason is here accorded equal status with experiment. Newton's theory is to be rationally ordered; as the method of composition requires, it will be set out in the Euclidean manner, as a deductively linked series of propositions in which the properties of light will be *proposed;* that done, its theorems can be experimentally checked, or *proved.* In fact only Book I of the *Opticks* is a complete science of this kind. Although Book II contains the analysis of various phenomena, it is not laid out as a deductive system; witness the fact that it contains *observations* rather than *propositions.* It is for that reason, I will argue, that Newton regards the conclusions of Book II as more dubitable than those of Book I.

For the time being, however, I want to linger on the words *proof* and *prove.* I have just glossed *proved* as "experimentally checked," and this is a long way from twentieth-century usage. For us, to prove a theorem is to show that the conclusion follows logically from the premises: formally, to show that the conclusion is in the deductive closure of the theory's axioms. A *proof by experiment,* on this account is almost a contradiction in terms. In contrast, at the end of the sixteenth century Marlowe wrote,

> Come live with me and be my love,
> And we will all the pleasures prove
> That hills and valleys, dales and fields,
> Or woods or steepy mountain yields.[7]

Clearly, as he used it, the word *prove* did not refer to anything recognizably Euclidean.

The three primary seventeenth-century significations of *to prove* are (1) "to experience," (2) "to test," and (3) "to establish the truth of" something. Even the last of these senses turns out to be some distance from ours. Marlowe used *prove* in the first—or possibly the second—sense. Nearer to Newton's time is a less ambiguous, if rather more plaintive example; a line by Cokaine from *The Tragedy of Ovid* (1662) runs, "I may prove the same sad destiny Clorinda did, should I become your wife" *(OED). To prove* here is surely just to experience.

According to the *OED* the second signification, where *to prove* is "to test," is "the prevailing use in the Bible of 1611" (the King James Version). A century later, in 1727, Hamilton coined the now proverbial phrase "The proof of the pudding is in eating it" *(OED),* but the classic example of this signification is another proverb, now more frequently misunderstood than not, "The exception proves the rule."[8] As an example of the third signification the *OED* cites Flavel in 1681: "A thousand witnesses cannot prove any point more clearly than the testimony of conscience doth." Note that this is still not the deductive sense of *prove;* for Flavel, the warrant of truth is not argument but inner experience.

In Newton's usage, I suggest, these three significations are interlocked: when a proposition is proved, it is tested and its truth confirmed by experiment. This is borne out by a striking phrase Newton uses in Proposition XI, Problem VI (p. 186); in a context like those in which he usually talks of making a *"proof* by experiment," he tells us to perform an experiment and "therein to *experience the truth of* the foregoing propositions" (emphasis added).

This view of Newton's "proofs" differs from the one proposed by Achinstein, who writes: "Whatever the validity or persuasiveness of the 'proofs by experiment' in the *Opticks,* I suggest that Newton treated them as 'deductions from phenomena.' He thought of them as demonstrating the truth of the proposition beyond reasonable doubt from facts themselves established by observing the results of experiments." I think, however, that this ignores the distinction, already remarked on, between analysis and synthesis, the two conceptually distinct phases of theory construction. The propositions that are "proved by experiment" are those appearing as theorems of the formally presented theory after synthesis has been effected. Some of these propositions will have been "deduced from phenomena" in analysis, but not all.[9] A proof by experiment, or confirmatory test, does not require the same range of data as an inductive generalization. Thus in the proof of Proposition I, Theorem I, it suffices to use just two colors, red and blue.

Another example, discussed by Achinstein in note 29 of his paper, is worth exploring in some detail. Proposition VI, Theorem V (p. 74) tells us that "the sine of incidence of every ray considered apart, is to its sine of refraction in a given ratio." That is to say, Snell's law, that $\sin i/\sin r$ is a constant, holds for all colors, not just for "rays which have a mean degree of refrangibility" (p. 76), and the value of the constant depends on the color in question.

Newton points out that, to show this, it is enough to show that, if we take a pair of colors, a and b, then, provided the angle of incidence is the same for both, $\sin r_a/\sin r_b$ is a constant for that pair. The details of his experimental proof need not concern us, but it involves passing light through a pair of prisms; the angles i, r_a, r_b, and so on are altered by using prisms of different angles. Possibly because each reading requires a different prism,[10] Newton performs only three experiments—hardly enough, one might think, on which to base an inductive generalization. Nor does he recommend that, ideally, more would be needed.

As we read on, however, it becomes clear that Newton is not engaged in a purely inductive science. In fact, he takes a very cavalier attitude to these particular experiments; he writes:

> The proportions of the sines being derived, they come out equal, so far as by viewing the spectrums, and using some mathematical reasoning I could estimate. For I did not make an accurate computation. So then the proposition holds true in every ray apart, so far as appears by experiment. And that it is accurately true, may be demonstrated upon this supposition. *That bodies refract light by acting upon its rays in lines perpendicular to their surfaces.* (p. 79)

Then follows a two-page argument from this "supposition" to the proposition, after which Newton concludes: "And this demonstration being general, without determining what light is, or by what kind of force it is refracted, or assuming anything further than that the refracting body acts upon the rays in lines perpendicular to its surface; I take to be a very convincing argument of the full truth of this proposition" (pp. 81–82).

This whole subsection of the *Opticks* is both very rich and very revealing. On the one hand, Newton appears as a master of the art of devising experimental techniques whereby the questions he wants to ask can be simply answered; on the other, although he says elsewhere that by doing such experiments we can "experience the truth" of propositions, in this instance he chooses to show that a proposition is "more accurately true" by deducing it mathematically from a premise of a particular kind. An added irony is provided by the fact that the premise (or "supposition") he uses is one that we would now reject, as it entails that the velocity of light increases when it passes into an optically denser medium. There is also an apparent discrepancy between his acceptance of this supposition and his avowed rejection of hypotheses.

To understand Newton here we have to go back to his early insistence that his "science of colors was mathematical."[11] This claim must be separated from the suggestion made in the *Opticks* (p. 404) that the proper methods of natural philosophy, the methods of analysis and of composition, are analogues of the methods of mathematics, from which indeed they take their names. This has often been thought a false analogy;[12] it is relevant here only because the method of composition yields a theory laid out *more geometrico*, and this is the sense in which Newton thought of his science as mathematical.

The method of composition displays the logical relations between a theory's propositions, and thus brings systematicity and unity to the theory: it shows that the truth of an individual proposition cannot be separately challenged. This mode of presentation is, however, possible only if the phenomena the theory deals with can be represented in a way that makes the methods of mathematical demonstration appropriate.[13] In the case of Newton's "science of colors," optical phenomena are at first modeled within geometry;[14] then, with the "supposition" of Proposition VI, Theorem V, this geometry becomes a geometrized kinematics. After putting forward his supposition, Newton introduces the "motions" of different rays: "But in order to this demonstration, I must distinguish the motions of every ray into two motions, the one perpendicular to the refracting surface, the other parallel to it, and concerning the perpendicular motion lay down the following proposition" (p. 79).

His argument involves a resolution (as we would say) of the velocity of the light into two components, one perpendicular to and one parallel to the refracting surface; according to the supposition, refraction alters only the component perpendicular to the surface. Although this is, as Newton acknowledges, a "supposition," it is clear that he regards it as a very weak one. It presupposes that light travels in rays with a finite velocity, but for Newton that would not have been problematic: first, the notion of a "light ray" is contained in the very geometric model that he uses, and second, he was aware that a value for the velocity of light had been obtained by Römer in 1673. Newton also assumes that, like any other motion, the motion of light can be resolved into components; since light falling perpendicularly onto an interface between two media suffers no refraction at the interface, it is natural to take the component of motion in that direction as having a privileged status. Now, any deviation involves a change in the ratio of this component of motion to the component parallel to the interface; Newton's supposition is that this change of ratio is due entirely to a change in one component, the component perpendicular to the interface.

In other words, the supposition goes very little beyond a purely descriptive representation of the phenomena in terms of what I have called geometrical kinematics; furthermore, it is expressed entirely in the vocabulary of that model. That is why Newton expects us to find it so persuasive. In particular, Newton draws attention to the fact that his demonstration is effected "without determining what light is, or by what kind of force it is refracted" (pp. 81–82). These words echo what he said thirty years earlier, when he was at pains to distinguish his theory from hypotheses about light; concerning his first presentation of the science of colors to the Royal Society, he wrote to Oldenburg:

I knew that the properties, which I declared to be of light, were in some measure capable of being explicated not only by [the corpuscular theory] but by many other mechanical hypotheses. And therefore I chose to decline them all and speak of light in general terms, considering it abstractedly as something or other propagated every way in straight lines from luminous bodies, without determining what that thing is.[15]

We read this, naturally enough, as an advocacy of a properly cautious epistemic strategy. But it also suggests a problem, which I will propose but not resolve: Could *any* theory about the nature of light be absorbed into Newton's natural philosophy without violating his self-imposed methodological constraints? For even if there were to be strong inductive evidence in favor of, say, the corpuscular hypothesis, that alone would not be enough; the hypothesis would have to be embedded within a deductive theory. And it is not clear how an account of the nature of light could *ever* be given in terms of the restricted set of concepts that Newton's finished theory employs.[16]

By a circuitous route I have arrived, finally, at Achinstein's "Newtonian argument" for the corpuscular theory of light. In what follows I will not be concerned with the merits of that theory, nor with those of the inductive argument advanced in its favor, but simply with the question whether or not Achinstein's argument is in accord with Newton's methodological views. Most of what I have to say can be inferred from what I have said already, so I will be brief.

Crucial to Achinstein's argument is the claim that, for Newton, the disjunction "Either light is a particle or a wave phenomenon" is true (or at least has a very high probability). This is a hard claim to make good, given Newton's explicit views on the topic: "I cannot think it effectual for determining truth to examine the several ways by which phenomena may be explained, unless where there can be a perfect enumeration of all those ways."[17] And, as we have seen, Newton took the properties of light to be explicable, not only by the corpuscular theory but "by many other mechanical hypotheses." Now, as Achinstein implies, there may be a conflict between what Newton says here and his third rule of reasoning in philosophy.[18] This rule allows us to extend inferences from observed members of a class to all other members, observable and unobservable alike, and Achinstein appeals to it in providing an inductive justification of the proposition "Light is either a particle or a wave phenomenon."

If the third rule does indeed license such an inference, then Newton is inconsistent. Before we convict him, however, we need to look again at this rule and its place in the *Principia*. As stated, the rule applies only to "qualities of bodies." But, as Achinstein points out, to hold Newton to this restriction would mean that he could not make any inductions at all concerning optical phenomena without assuming at the outset that light consists of "bodies." It is clear that the restriction needs to be relaxed; the question is, should it be relaxed so far that it allows inductive generalizations concerning causal processes? Surely not: for Newton, the third rule is important precisely because it allows him to avoid talking about such processes. It is the third rule that justifies him in taking the law of universal gravitation to be true, even though he has no adequate causal explanation for it.

My view is that the restriction on inductions that meshes best with Newton's practice is one that limits induction to those features of the world that are mathematically representable. Certainly the only inductive generalizations that can appear in a finished theory are of this kind; it is also true that, in the *Principia*, the relevant features of the world are the ones he mentions, the quantifiable properties of bodies.[19]

This proposal, however, may well rule out the induction about types of causal process that Achinstein suggests. In making it, I am harping on a now familiar theme. For my greatest reservation about Achinstein's suggestion is that, by focusing as he does on inductive justification, he ignores Newton's specification of the form a scientific theory should take.

Descartes and Newton both took it for granted that science should be presented within a deductive system. They disagreed about the scope of that system and about the way its axioms were to be justified, and it is these differences that are stressed in Newton's discussions of methodology. Thus, at the end of the *Opticks* Newton is at pains to explain and to justify the role of induction in his science; hence the comparatively detailed treatment of the method of analysis in these pages. Synthesis, in contrast, required for Newton's audience neither justification nor explanation, and so could be dealt with in a mere four lines (p. 405). But this does not mean that it was unimportant. When Newton wrote (p. 404) that in "the investigation of difficult things the method of analysis ought ever to precede the method of composition," he was not advocating a science that contained nothing but inductive generalizations. The science he sought, and to a great degree achieved, was an integrated body of knowledge, a true partnership of Reason and Experiment.

Notes

1. Sir Isaac Newton, *Opticks*, ed. I. B. Cohen (New York: Dover, 1952); page numbers in the text will refer to this edition.

2. On this question McGuire writes, "Newton did not see any dichotomy in method between that used in the *Principia* and that found in the *Opticks*." J. E. McGuire, "Newton's Principles of Philosophy: An Intended Preface for the 1704 *Opticks* and a Related Draft Fragment," *British Journal for the Philosophy of Science* 5 (1970), 178–186. Which is not, of course, to say that differences do not exist.

3. McGuire, 1970.

4. Sir Isaac Newton, *Philosophiae Naturalis Principia Mathematica*, ed. F. Cajori (Berkeley: University of California Press, 1934).

5. Recall that for Newton the qualification expressed by "perhaps" was significant. Letter to Oldenburg, June 11, 1672, in H. W. Trumbull, ed., *The Correspondence of Isaac Newton*, vol. 1 (Cambridge: Cambridge University Press, 1959), pp. 173–174.

6. McGuire, 1970, p. 185.

7. C. Marlowe, "The Passionate Shepherd to His Love." Howard Stein suggested to me that Donne's lines "Come live with me and be my love, / And we will some new pleasures prove" contain the implicit criticism that Marlowe based his seductive generalization on too narrow a sample.

8. I owe this example to Ronald de Sousa.

9. Note that the change of signification of *deduce* since 1700 is parallel to that of *prove*.

10. Thus on p. 78 he describes how water prisms may be constructed to make up for a lack of glass ones.

11. Letter to Oldenburg, June 11, 1672, in Trumbull, 1959, p. 187. Here and elsewhere I modernize the spelling.

12. For a discussion of such criticisms, see McGuire, 1970, p. 185, n. 14.

13. See my introduction to this volume, and also E. A. Burtt, *The Metaphysical Foundations of Modern Science*, 2d ed. (Garden City, N.Y.: Doubleday, 1954), pp. 220–226.

14. Note in passing that Book I of the *Opticks* violates the canons of strict deduction just where Newton moves away from geometry to introduce the phenomenological distinction between colors (p. 20).

15. Letter to Oldenburg, June 11, 1672, in Trumbull, 1959, p. 174.

16. I postpone a discussion of this problem to a future occasion.

17. Letter to Oldenburg, July 6, 1672, in Trumbull, 1959, p. 209.

18. *Principia*, pp. 398–400.

19. This claim would need to be spelled out more carefully than the length of this paper permits.

Chapter 10

Kant and Newton: Why Gravity Is Essential to Matter

Michael Friedman

This year we honor the three hundredth anniversary of the publication of Isaac Newton's widely and justly celebrated *Mathematical Principles of Natural Philosophy.*[1] Last year we observed the two hundredth anniversary of the publication of one of the relatively less well known works of Immanuel Kant: his *Metaphysical Foundations of Natural Science.*[2] The latter work, although it has not been widely studied—at least in the English-speaking world—in fact plays a central role in the development of Kant's critical philosophy. It was written between the first and second editions of the *Critique of Pure Reason,* and accordingly it helps to explain some of the significant shifts in emphasis that are found in the second edition.[3] In particular, the second edition emphasizes the importance of space and the fact that Kant's basic objects—namely "appearances"—are physical bodies located outside us in space. This in turn fits in with Kant's concern to answer the charge—made in a well-known hostile review of the first edition of the *Critique of Pure Reason* by Garve and Feder—that the critical philosophy is really just a variant of "subjective" or Berkeleyan idealism.[4] As I understand Kant's answer—in the *Prolegomena,* the *Metaphysical Foundations,* and the second edition of the *Critique*—the point is that the critical philosophy is not a system of "phenomenalism" or "subjective idealism" at all, but rather a system of what we might call transcendental physics.

This becomes especially clear in the *Metaphysical Foundations,* where Kant argues that the metaphysical doctrine of body, the doctrine of spatial objects in outer sense, is the *only* realization or instantiation of the transcendental metaphysics of the first *Critique:* a realization without which the transcendental principles of the first *Critique* would be empty of all sense and meaning (preface: p. 16). Moreover, when one looks at the content of the *Metaphysical Foundations,* it turns out that the fundamental principles of this metaphysical doctrine of body are basically Newton's laws of motion (in the third chapter, or Mechanics) and that the primary application of these principles is to Newton's theory of universal gravitation (in Props. V–VIII of the second chapter, or Dynamics). In this way, the mathematical physics of New-ton's *Principia* turns out to be the fundamental exemplar of Kant's transcendental philosophy as applied *in concreto*. At the same time, however, Kant's *Metaphysical Foundations* is also an extremely important contribution to the critical analysis of the conceptual foundations of Newtonian physics—or so I shall argue.

Kant's one explicit criticism of Newton in the *Metaphysical Foundations* occurs in Observation 2 to Proposition VII of the second chapter, or Dynamics. Kant quotes Newton's well-known words from the Advertisement to the second edition of the *Opticks:* "And to show that I do not take Gravity for an essential Property of Bodies, I have added one Question concerning its Cause," where the reference is of course to Query 21.[5] Kant takes issue with this famous Newtonian denial and argues that if Newton does not conceive gravity as an essential property of matter, he is necessarily "set at variance with himself" (p. 66). Similarly, Kant disputes the "common opinion" that "Newton did not find it necessary for his system to assume an immediate attraction of matter; but rather, with the most rigorous abstinence of pure mathematics, he left the physicists full freedom to explain the possibility of attraction as they saw fit—without mixing his propositions with their play of hypotheses" (p. 64). According to Kant, Newton may not leave open such possibilities—for example, the possibility of explaining gravitational attraction by an "Aetherial Medium" as in Query 21; on the contrary, in the words of Kant's own Proposition VII: "The *attraction essential to all matter* is an immediate action of one matter on another across empty space."

Kant first argues that gravitational attraction is essential to matter in Proposition V of the Dynamics, which reads: "The possibility of matter requires a *force of attraction* as the second essential fundamental force of matter." Moreover, the conclusion of the proof of this proposition clearly states that "an original attraction belongs to all matter, as a fundamental force belonging to its essence" (p. 57). Thus, there is no doubt at all that Kant himself thinks that gravity is essential to matter. When we examine the proof of this proposition that Kant offers here, however, we are liable to be disappointed. The proof proceeds by a "balancing" argument: if matter were endowed only with repulsive force, it would disperse to infinity, and then space would in fact be empty of matter; hence the possibility of matter requires an attractive force in order to set limits on the expansion inevitably following from a repulsive force. Conversely, as Kant argues in Proposition VI, if matter were endowed only with attractive force, it would contract to a point, and space would once again in fact be empty of matter; therefore both attractive and repulsive forces belong to the essence of matter (note to Prop. VI).

This argument is disappointing for the following reasons. First, it is not immediately clear why the repulsive force assumed in the proof of Proposition V must be limited by a fundamental attractive force. Why can limits not be set by another repulsive force, expanding in the contrary direction, as it were? Second, even if we grant the inference from repulsion to attraction, the argument goes through only if we begin by assuming a fundamental force of repulsion. Newton himself is committed to no such repulsion, however; on the contrary, he is committed to the "absolute impenetrability" that Kant explicitly opposes to the "relative impenetrability" effected by a fundamental force of repulsion (Def. 4 of the Dynamics). Newton, that is, is committed to the "solid, massy, impenetrable, moveable Particles" of Query 31 of the *Opticks*.[6] Yet Kant claims that in rejecting gravity as essential to matter, Newton is "set at variance with himself," and it is not at all clear how the present argument can show this. Finally, even if we waive both of the above problems, it is hard to understand the place of such an argument in Kant's critical philosophy. For it appears to be a physical argument—or, what is worse, a metaphysical argument in just the sense in which Kant the critical philosopher has promised to eschew metaphysics: that is, the argument proceeds by considering the possible behavior of matter as it is or may be "in itself," quite independently of any relation of such matter to our cognitive faculties. Kant clearly insists in the preface to the *Metaphysical Foundations*, however, that "all true metaphysics is drawn from the essence of the thinking faculty itself" (p. 9). It is necessary, then, to look for a more subtle interpretation of Kant's argument.

The first point to notice is that the "balancing" argument of Proposition V is not meant to be an explanation of how matter acquires a determinate volume—that is, how a particular piece of matter acquires a definite exterior surface and thus becomes a body with a determinate shape. For Kant repeatedly emphasizes that the explanation of *cohesion* lies entirely outside of the *Metaphysical Foundations*, as a physical rather than a metaphysical problem (see note 2 to Prop. VIII of the Dynamics; General Observation to Chapter 4, or Phenomenology). Accordingly, Kant sharply distinguishes between the concept of *matter* in general and the more determinate—and physical— concept of a *body* (§§1, 2 of the General Observation to Dynamics). What the "balancing" argument of Proposition V is supposed to explain, then, is not the possibility of bodies—of particular pieces of matter with definite volumes and figures—but the possibility of matter in general, whether or not it is confined to a definite volume. Specifically, what has to be explained is how "matter fills a space with a determinate degree" (proof of Prop. VII of the Dynamics, note 1 to Prop. VIII): in other words, as we would now put it, the problem is to explain how matter has a determinate *density*. This is the same problem, however, as explaining how matter can have a determinate *mass* or *quantity of matter* (Obs. 1 to Prop. IV, Proof of Prop. V). And again, Kant emphasizes that the only "a priori comprehensible universal characteristics of matter" are *elasticity* (owing to repulsive force) and *weight* (owing to attractive force and proportional to mass); *cohesion* is not itself such a characteristic (note 2 to Prop. VIII).

But what kind of explanation of the possibility of matter having a determinate mass or quantity of matter is the "balancing" argument supposed to provide? In complete agreement with the above distinction between metaphysical and physical levels, Kant explicitly argues that a mathematical-physical explanation (in his language a "construction") of the possibility of matter filling space in a determinate degree does not belong to metaphysics (note 1 to Prop. VIII). He speculates on how such a "construction" might go (by considering the laws of attractive and repulsive forces and their interplay, Obs. 1 to Prop. VIII), but clearly asserts that "metaphysics is not responsible if the attempt to construct matter in this way should perhaps not succeed" (p. 69). What Kant has provided, then, is not a mathematical-physical explanation but a metaphysical one; and, I suggest, this is to be understood as a metaphysical explanation in Kant's own peculiar sense of metaphysics, which is to be "drawn from the essence of the thinking faculty itself." In other words, when Kant here explains the possibility of matter, this is to be understood as an explanation of how matter is possible *as an object of knowledge and experience*. Specifically, to explain how a determinate mass or quantity of matter is possible here means to explain how we can cognize or estimate the masses of objects. This suggestion is consistent with Kant's language in the proof of Proposition V, where he argues that without a fundamental force of attraction, no "assignable [*anzugebende*] quantity of matter would be found" (p. 57).

This suggestion also allows us to connect Proposition V with Kant's explicit criticism of Newton in the Observation to Proposition VII. For Kant's specific argument there is that if Newton rejects gravity as essential to matter, he is left with no way of "grounding" the proposition that gravitational attraction is directly proportional to mass. If Newton leaves open the possibility of an explanation of gravity in terms of an "Aetherial Medium," then, Kant asks (pp. 64–65): "How could he ground the proposition that the gravitational attraction of bodies that they exert at equal distances around themselves is to be proportional to the quantity of their matter, if he did not assume that all matter—merely as matter and through its essential property— exerts this moving force?" What Kant has in mind here, I think, is Newton's use of universal gravitation—as a force whose magnitude is directly proportional to mass or quantity of matter—in actually estimating or determining the masses of the primary bodies in the solar system.

Newton employs universal gravitation in this way in the Corollaries to Proposition VIII of Book III of *Principia*. The idea is to determine the relative masses of the sun, Jupiter, Saturn, and the earth by comparing the *accelerations* of the satellites of these primary bodies: specifically the accelerations of Venus, a moon of Jupiter, a moon of Saturn, and the earth's moon. According to the law of gravitation, these accelerations are proportional, at equal distances from their respective primary bodies, to the masses of these primary bodies. Moreover, these accelerations can themselves be directly computed from the distances and periodic times of the satellites. Finally, using the inverse-square law, we can compare these accelerations at equal distances. The result is a determination of the masses of the sun, Jupiter, Saturn, and the earth: if the sun is assigned unit mass, then Jupiter's mass is as 1/1,067, Saturn's is as 1/3,021, and the earth's is as 1/169,282 (Cor. II).[7]

That Kant in fact has this Newtonian procedure in mind here receives strong confirmation from the conclusion of his explicit criticism of Newton. If Newton denies that gravitation is essential to matter, then, according to Kant: "He could absolutely not say that the attractive forces of two planets, e.g., that of Jupiter and Saturn, which they manifest on their satellites (whose masses one does not know) at equal distances, relate to one another as the quantity of matter of these heavenly bodies, if he did not assume that they, simply as matter and therefore according to a universal property of matter, would attract other matter" (p. 66). Kant's argument, then, appears to go as follows: universal gravitation is necessarily employed in estimating or determining the masses of bodies; a fundamental attractive force is therefore necessary for matter to have an assignable quantity of matter as an object of experience; therefore a fundamental attractive force is essential to the possibility of matter as an object of experience.

This argument contains an important insight into Newton's actual procedure or methodology in the *Principia*, for it highlights the crucial role played by universal gravitation in estimating the masses of the bodies in the solar system. Nevertheless, the argument is still quite insufficient for Kant's purposes. Kant wants to show that gravitation is essential to matter: that it is an "original fundamental force" which cannot itself be explained by the action of repulsive forces—by the pressure exerted by an "aether," for example. Newton's procedure, however, appeals only to basic mathematical facts about gravitational attraction, in particular to the inverse-square law and proportionality to mass. Why should these facts not themselves be explained in more basic terms—by Newton's "Aetherial Medium" of Query 21, say? Second, and even more fundamentally, Kant wants to show that gravity is essential to matter as such—to matter as an object of experience. Even if we are able to show that gravitation, whether explained in more basic terms or not, is essential for matter to have an assignable mass, it is not yet clear why *this* is essential to the possibility of matter as such. Why can we not conceive matter as merely extended, impenetrable, and movable, for example?

To answer these questions we need to go a bit more deeply into Kant's critical analysis of the conceptual foundations of Newtonian physics. At the most fundamental level, this analysis is based on a rejection of absolute space, which, according to Kant, lies beyond the bounds of all possible experience. Yet Kant also clearly recognizes that Newton's physics requires a distinction between absolute and relative motion or, in Kant's own terms, a distinction between "true" and "apparent" motion. Such a distinction is certainly required, for example, if we are going to claim, following Newton, that the Copernican system is much closer to the truth than the Ptolemaic system.[8] For Kant, however, the distinction between true and apparent motion cannot be drawn on the basis of a preexisting absolute space; rather, it must invoke only materials that are themselves drawn from our actual experience of nature: our concern is always with "the movable, in so far as it, as such a thing, can be an object of experience" (Def. to the fourth chapter, or Phenomenology, of *Metaphysical Foundations*).

Kant defines matter as "the movable in space" (Def. 1 to the first chapter, or Phoronomy); and for him both the space and the motion in question must be determinable, in principle, by means of experience itself. How is this to be done? Here Kant sees Newton's procedure for determining a privileged frame of reference (what we would now call an *inertial* frame of reference), as this procedure is concretely executed in Book III of *Principia*, as what can alone give objective meaning to the idea of absolute space and the consequent distinction between true and apparent (absolute and relative) motion. This procedure does not find or discover a preexisting absolute space, as it were; rather, it alone first gives meaning and content to the idea of absolute space itself. Thus, what Newton does in Book III is to employ his laws of motion, especially the first and third laws, so as to determine a privileged frame of reference fixed at the center of mass of the solar system and in which the fixed stars are at rest. This frame turns out to be centered sometimes within, sometimes without, the surface of the sun, but never very far from the sun's center (Prop. XII of *Principia*); and this is the precise sense in which the Copernican system is closest to the truth. In the absense of the Newtonian procedure for determining this center of mass frame, however, not only would the issue between Copernicus and Ptolemy remain undecided, but it would, strictly speaking, lack all objective meaning.

We are now in a position to understand Kant's answer to the second of the two questions just raised. Matter is the movable in space. But to give objective meaning in experience to the idea of motion— that is, to the distinction between true and apparent motion—we need a procedure for determining, in experience, a privileged frame of reference. This procedure—Newton's procedure—presupposes the laws of motion; and this, by the way, is why Kant views the laws of motion as themselves a priori: they do not state facts about motion, as it were, but make the very idea of motion first possible. Moreover, this procedure also presupposes a means for actually estimating or determining the masses of the bodies in question, for otherwise we are unable to determine the relevant center of mass. As we have already seen, however, the way this is actually done in Book III of *Principia* is by means of universal gravitation: this is how the masses of the primary bodies in the solar system are actually estimated, and this, therefore, is what enables us to conclude that the center of mass of the solar system is very close to the center of the sun. Thus, in giving objective meaning in experience to matter as the movable in space, we presuppose universal gravitation as well; and this is the sense in which the latter, for Kant, is necessary for the possibility of matter as such.

So much for Kant's answer to our second question. The really difficult and interesting problem is raised by our first question, however. Granting that universal gravitation, as proportional to mass and as governed by the inverse-square law, is necessary for estimating the masses of the relevant bodies—and therefore for determining a privileged center of mass frame—why must we conceive this attraction as itself *essential* to matter? Why can we not leave open the possibility that gravitational attraction, together with its laws, may eventually be explained in more basic terms, such as the pressure exerted by an aether? As far as the abstract truth of the law of gravitation is concerned, Kant must admit, I think, that such a more basic explanation is indeed possible. As always, however, his concern is not with such abstract truth but rather with the question of our knowledge or possible cognition of the truth. His question concerns precisely how the law of gravitation—and especially the fact that gravitational attraction is proportional to mass—is itself "grounded" for us. His claim, I think, is that this last fact cannot be epistemically grounded unless we presuppose gravitational attraction as essential to matter: that is, as acting *immediately*, with no need whatever of an intervening medium, and as *universal*.

Here we must carefully distinguish two different aspects of the fact that gravitational attraction is proportional to mass. The first concerns the mass of the body being attracted: a satellite of a primary body, for example. The *acceleration* of such a satellite toward the primary body depends only on its distance from the primary body and is entirely independent of the mass of the satellite (all bodies fall the same in a gravitational field). It follows that the gravitational *force* acting on the satellite is, at a given distance from the primary body, directly proportional to the satellite's own mass: the mass of the satellite, in other words, is directly proportional to its *weight*. This property of gravitational attraction follows from Kepler's third law, which gives the period, and therefore the acceleration, of any satellite solely as a function of its distance from the primary body (from Kepler's third law it of course also follows that this acceleration is proportional to the inverse square of the distance). Moreover, this property of gravitational attraction is so derived, in conjunction with the "moon test" and the observed proportionality between mass and weight in the action of terrestrial gravity, in Proposition VI of Book III of *Principia*.

There is, however, a second property of gravitational attraction which is quite distinct from this first one. Not only is the acceleration of a satellite toward its primary body independent of the mass of the satellite, but this same acceleration is dependent on the mass of the primary body: more precisely, at a given distance from the primary body this acceleration (of the satellite) is directly proportional to the primary body's own mass. This property does not immediately follow from the first one; and our first property, by itself, gives us no means whatever for universally comparing the masses of the various primary bodies. It follows, then, that it is this second property of gravitational attraction which is crucially important for comparing the masses of the different primary bodies in the solar system in Proposition VIII.

Newton derives this second property of gravitational attraction in Proposition VII, which itself depends on Proposition LXIX of Book I *and* on our first property of gravitational attraction. The idea can be illustrated by considering the particular comparison of the masses of

Jupiter and Saturn (this, you will recall, is the example Kant himself fastens upon). We know, by our first property, that both Jupiter and Saturn are surrounded by inverse-square "acceleration-fields": that is, the satellites of both primary bodies experience inverse-square accelerations that are entirely independent of the masses of the satellites.[9] We want to use these accelerations, at a given distance from the primary bodies, as measures of the masses of the primary bodies themselves. How do we establish this proportionality? We assume that the inverse-square "acceleration-fields" of the two planets extend far beyond the regions of their respective satellite systems and, in fact, that they extend to the two planets themselves: there is an inverse-square acceleration of Saturn toward Jupiter and an inverse-square acceleration of Jupiter toward Saturn. We then apply Newton's third law of motion—the equality of action and reaction—to these last two accelerations: the mass of Jupiter times the acceleration of Jupiter toward Saturn is equal to the mass of Saturn times the acceleration of Saturn toward Jupiter. It follows that the ratio of the two accelerations is equal to the ratio of the two respective masses; and therefore, since these "acceleration-fields" are assumed to be identical to those acting on the *satellites* of Jupiter and Saturn, it follows that the accelerations of these satellites are, at a given distance, themselves proportional to the masses of the two primary bodies.[10]

What matters here are two general features of the argument. First, we need to extend the "acceleration-fields" of the two planets far beyond the regions of their respective satellite systems and suppose that they affect the two planets as well. In other words, we need to assume that gravitational attraction takes place not just between primary bodies and their satellites but also between the primary bodies themselves. This is a first, and absolutely crucial, step on the way toward *universal* gravitational attraction. Second, we need to apply Newton's third law of motion—the equality of action and reaction—to the gravitational interaction between the primary bodies, in our case, to the interaction between Jupiter and Saturn. For only so do the masses of the primary bodies enter directly into our calculations at all. And these masses, of course, are precisely what we have been aiming at all along.

Suppose now that we seriously entertain the possibility of an explanation of gravitational attraction by the action of an "Aetherial Medium." This would cause overwhelming difficulties for our derivation. First, on this kind of model of gravitational attraction one would have no reason to extend the attractions in question beyond the regions of the respective satellite systems. Indeed, one would naturally expect the attractions to be limited to precisely these regions; and, in any case, it would be extremely difficult to incorporate *universal* attraction coherently into the model. Second, and even more fundamentally, however, one would have no license whatever for applying the third law of motion—applying the equality of action and reaction to the accelerations of Jupiter and Saturn, for example. For, on an aether model, even if one could somehow contrive to incorporate accelerations of Jupiter and Saturn toward each other, there would be absolutely no warrant for applying the equality of action and reaction directly to these two accelerations themselves: on the contrary, this principle is to be applied to the interactions among Jupiter, Saturn, *and the intervening medium.* Conservation of momentum will not hold, in other words, if we consider merely the accelerations of Jupiter and Saturn alone.

This point is in fact explicitly put to Newton in a well-known letter (of February 18, 1712) by Roger Cotes, a most acute Newtonian who was of course himself strongly tempted to conceive gravity as an essential property of matter.[11] Cotes argues that the third law of motion can be applied only to cases of "Attraction properly so called," but not in cases of merely apparent attraction affected by pressure (Cotes illustrates the point by considering one globe rotating about another on a table in virtue of being pushed by an "invisible Hand"). Cotes even remarks that neglect of this point causes difficulties with the argument of Proposition VII. Kant, as I read him, is entirely in agreement with this fundamental observation of Cotes. Kant goes further, however, by also stressing the importance of the assumption of universality in the argument of Proposition VII. We must presuppose, according to Kant, that the action of gravitational attraction is *immediate:* effected by no intervening medium. We must also presuppose that the action of gravitational attraction is *universal:* acting between each body in the solar system and each other body. Together, these two presuppositions constitute the content of Kant's Propositions VII and VIII of his Dynamics; and the two together amount, in Kant's eyes, to the claim that gravitational attraction is an essential property of matter.

It is worth reminding ourselves of precisely what this claim means for Kant. We need to presuppose the immediacy and universality of gravitational attraction in order to develop a rigorous method for comparing the masses of the primary bodies in the solar system. We need such a method, in turn, in order rigorously to determine the center of mass of the solar system. This, in turn, is necessary for rigorously determining a privileged frame of reference and thus for giving objective meaning, in experience, to the distinction between true and apparent (absolute and relative) motion. This, finally, is necessary if matter, as the movable in space, is to be itself possible as an object of experience. Hence an essential—that is, immediate and universal—attraction is necessary to matter as an object of experience. It follows, for Kant, that the immediacy and universality of gravitational attraction must be viewed, like the laws of motion themselves, as in an important sense a priori. These two properties cannot be straightforwardly obtained from our experience of matter and its motions—by some sort of inductive argument, say—for they are necessarily presupposed in making an objective experience of matter and its motions possible in the first place.

Now this last Kantian claim—that the immediacy and universality of gravitational attraction must be viewed as in an important sense a priori—may seem obviously and outrageously false; and this is especially so in the case of the universality of gravitation. Is this property not subject to inductive confirmation of the most straightforward kind: namely, by observations of the planetary perturbations? Does not Newton, in the third edition of *Principia*, appeal to observations of precisely such perturbations in the orbits of Jupiter and Saturn? Does not Kant's correspondent J. H. Lambert himself make an important contribution in 1773–1776 to the theory of the so-called great inequality of Jupiter and Saturn, by providing tables—obtained purely empirically—that suggest for the first time that the anomaly in question is periodic?[12]

These facts are undeniable, and they do show that the assumption of universality has indeed been corroborated by observational data. Kant's problem, however, concerns the interpretation of such observations. For what do such observational data—Lambert's, for example—actually show? They show that in a given frame of reference, perhaps a frame of reference fixed at the center of the sun and in which the fixed stars are at rest, certain motions that we call planetary perturbations occur. Yet such data, by themselves, do not and cannot show that the motions in question can be interpreted as true or absolute motions: that *absolute* gravitational acceleration is universal. Kant's point is that this last notion of true or absolute motion does not even have objective meaning or content unless we employ Newton's procedure for determining the center of mass of the solar system and hence presuppose that absolute gravitational acceleration is in fact universal. To be sure, observations of planetary perturbations turn out, fortunately, to provide corroboration for this whole scheme; but consider what our situation would have been if such perturbations had not been observed. We would not then merely be in the position of having disconfirmed an empirical hypothesis, the hypothesis that gravitational accelerations are universal; rather, we would be left with no coherent notion of true or absolute motion at all. For the spatio-temporal framework of Newtonian theory—which, for Kant, can alone make such an objective notion of absolute motion first possible—would itself lack all objective meaning.

It is along these lines, then, that I propose to interpret Kant's claim that gravitational attraction is essential to matter. Two problems remain, however. The first is more narrowly textual: our interpretation is based largely on Kant's observations concerning the Newtonian procedure for measuring or estimating the masses of the primary bodies in the solar system, as expressed in Propositions VI and VIII of the Dynamics chapter of Kant's *Metaphysical Foundations;* but how do these considerations relate to the earlier "balancing" argument by which Kant attempts to deduce the necessity of a fundamental force of attraction in Proposition V? The second problem concerns the broader philosophical significance of Kant's claim: What, in particular, are we to make of the apparently flat-out contradiction between this claim and the views of Newton himself?

With respect to the first problem, it must be admitted that our considerations do not appear to be directly related to the earlier "balancing" argument. There may nonetheless be an indirect connection, however; and, in fact, I think that just such a connection may be discerned in Kant's Observation to his Proposition V. In this Observation Kant steps back from the "balancing" argument itself and reflects on his procedure. Since repulsion and attraction are both essential to matter, Kant asks, why do we postulate the first and derive the second only through "inferences"? Why do we not simply postulate both fundamental forces on the same level? The reason, according to Kant, does not depend on the fact that repulsive force is more directly perceived by our sense organs; for this is a merely contingent and physiological fact about our empirical constitution. The true reason, rather, is to be sought in the differing properties of the two fundamental forces themselves. In particular, the fundamental force of attraction of a body such as the earth is exerted precisely as if all the matter of the body were concentrated in a single point located at its center of mass. This force by itself cannot therefore provide us with information about the volume, figure, or even the position of such a body; for in acting on our sense organs, whatever their constitution may be, it actually provides us with only a direction in space. We need to start with repulsive force, consequently, which alone can provide us with volume, figure, and position—and which thus provides the basis for the "first application of our concepts of *magnitude* to matter" (p. 58).

How are we to understand this passage? Note first that, in appealing to the fact that the gravitational force of a body such as the earth can be conceived as exerted by a single point concentrated at its center of mass, Kant is alluding to Proposition VIII of Book III of *Principia:* the very proposition that is central to our considerations about measuring or estimating the masses of the heavenly bodies. What Kant is saying, as I read him, is that, although a system of mere mass points could therefore realize—provide a model for, as it were—Newton's theory of the solar system, a system of mere mass points could not be cognized as such by us. For we need first to observe or cognize the figures and especially the relative positions of the heavenly bodies in order to derive, by "inferences," the Newtonian theory of gravitation from such observations. We do this by means of the *light rays* that proceed from the sun and reflect off the heavenly bodies to our eyes, and this in turn depends on the impenetrability of the heavenly bodies and thus, for Kant, on their repulsive force.[13]

The "balancing" argument can therefore be reinterpreted from an epistemological point of view as follows: Without repulsive force—with only attractive force, in other words—we would be unable to cognize the volumes, figures, and relative positions of the objects of our theory. Each heavenly body would, from an epistemological point of view, "contract to a point"; and, although such a system of mere mass points could certainly provide a model for our theory, it could never be cognized as such by us. Hence, we need to start with a fundamental force of repulsion. By contrast, if we had only such a fundamental force of repulsion, we would have no means for cognizing or determining the relative masses of the heavenly bodies, and thus in the end no means for coherently assigning these bodies determinate states of true or absolute motion. Without attractive force, in the words of Kant's proof of Proposition V, "no assignable quantity of matter would be found in any assignable space" (p. 57).

I turn now to the problem of the broader philosophical significance of Kant's claim. What are we to make of the fact that Newton, the great inventor of the theory of gravitation, explicitly and firmly denies that gravity is essential to matter, whereas Kant—who I hope I have convinced the reader has a claim to be considered the great philosophical interpreter of Newton's theory—just as explicitly and firmly asserts this? Do we here have simply a flat-out contradiction between the opinions of these two great thinkers? Must one therefore be simply right and the other then simply wrong? This does not seem to me to be a profitable way to view the situation.

First, when Newton denies that gravity is essential to matter and Kant affirms this, it is clear that the two thinkers do not at all mean the same thing by "essential." Kant, as we have seen, is making basically an epistemological point. Gravity, as postulated to act immediately and universally, plays an essential role in Newton's procedure for measuring or estimating the masses of the heavenly bodies and thus in his construction of a privileged frame of reference: the center of mass frame of the solar system. This in turn, for Kant, is a necessary condition for providing the notion of true or absolute motion with objective meaning and, in particular, for making an unambiguous determination of this notion with respect to the observed, merely relative, or apparent motions in the solar system. Hence, Kant concludes, a fundamental force of attraction—that is, gravity—is essential to matter, defined as the movable in space, as an object of knowledge or experience.

Kant's claim, then, is to be understood in the context of his transformation and reinterpretation of metaphysical questions generally. The older tradition understands metaphysical questions to be entirely distinct from epistemological ones. In Kant's terms, this tradition is concerned with the nature of objects as they are or may be "in themselves"—entirely independent of us and our cognitive faculties—and, in particular, with the relation of such objects to the Divine Power. Thus, for example, Descartes notoriously argues that mind and body are essentially distinct (distinct "in nature") by arguing that the Divine Power can separate them. Kant, by contrast, in arguing that gravity is essential to matter, has broken decisively with any concern whatever about the nature of matter "in itself" and, in particular, with any concern about the relation of matter to the Divine Power. Metaphysics, at least insofar as this science is concerned with theoretical questions involving truth and falsity, must henceforth relinquish all such concerns and instead take as its subject matter the relation of objects (in this case matter) to our cognitive faculties. (This idea, of course, is what Kant himself calls the Copernican revolution in metaphysics.) What is most noteworthy and remarkable about this Kantian transformation of the very meaning and subject matter of metaphysics, in the present context, is that it is largely effected by reflecting on *Newton's* methodological practice, as exemplified above all in the *Principia*.

Yet, from Kant's point of view, Newton's metaphysical pronouncements (in this case the denial that gravity is essential to matter) have not kept pace with his own methodological practice. For Newton's conception of the meaning and subject matter of metaphysics, it is clear, is precisely that of the older tradition Kant is attempting to transform. In denying that gravity is essential to matter, in particular, Newton is expressing a conviction about the nature of matter "in itself" and, specifically, about its possible relation to the Divine Power. Thus, in a well-known letter to Richard Bentley of February 25, 1692, Newton says: "It is inconceivable, that inanimate brute Matter should, without the Mediation of something else, which is not material, operate upon, and affect other matter without mutual Contact, as it must be, if Gravitation in the Sense of *Epicurus*, be essential and inherent in it. And this is one Reason why I desired you would not ascribe innate Gravity to me."[14] This conviction is plainly of an entirely different kind from the methodological considerations we have seen driving Kant to the opposite conclusion: it is of a piece with such other well-known Newtonian convictions as that (as Newton expresses it in the same exchange with Bentley) the orderly structure of the solar system—the "Frame of the World"—could never arise from a "Chaos of Matter" alone without the intervention of the Divine Power.[15]

The passage from the letter to Bentley continues as follows (and I hope I will be forgiven for quoting these well-known words once again):

> That Gravity should be innate, inherent and essential to matter, so that one Body may act upon another at a Distance through a *Vacuum*, without the Mediation of any thing else, by and through which their Action and Force may be conveyed from one to another, is to me so great an Absurdity, that I believe no Man who has in philosophical Matters a competent Faculty of thinking, can ever fall into it. Gravity must be caused by an Agent acting constantly according to certain Laws; but whether this Agent be material or immaterial, I have left to the Consideration of my Readers.[16]

Newton does not of course explicitly assert that the "Agent" responsible for gravity is either immaterial or divine. Nevertheless, although we are clearly on shaky ground here, it is perhaps permissible to speculate that this is in fact his true conviction. Such a claim is explicitly made by Bentley, in his *Confutation of Atheism from the Origin and Frame of the World*—the work Newton is here commenting on—and Newton never attempts to correct Bentley on this score.[17] Moreover, this could also explain why Newton never takes seriously the problem about conservation of momentum and the third law of motion raised by the acute Professor Cotes (for the actions of an *immaterial* medium are of course not subject to the third law). The idea that gravity is effected by an immaterial medium through the Divine Power is also consistent with such passages as that from Query 31 of the *Opticks*, where Newton speaks of "a powerful ever-living Agent, who being in all Places, is more able by his Will to move the Bodies within his boundless uniform Sensorium, and thereby to form and reform the Parts of the Universe, than we are by our Will to move the Parts of our own Bodies."[18]

In any case, it is well worth noting, finally, that Kant, despite his separation of the question whether gravity is essential to matter from all considerations involving the Divine Power, manages to find a reflection—or, perhaps better, an analogue—of such a Newtonian conception of the relationship between space, gravity, and divinity in the completed system of his critical philosophy. This comes about in the following way.

Kant, as we have seen, begins by rejecting Newtonian absolute space. In its place he puts a methodological procedure for determining a privileged frame of reference, a procedure that he derives from the argument of *Principia*, Book III. There, as we know, Newton's argument culminates in the determination of the center of mass frame of the solar system, relative to which neither the earth nor the sun can be taken to be exactly at rest. For Kant, however, our methodological procedure cannot, strictly speaking, terminate at this point; for the center of mass of the solar system itself experiences a slow rotation with respect to the center of mass of the Milky Way galaxy. Nor can even this last point furnish us with a privileged state of rest; for, according to Kant, the Milky Way galaxy also experiences a rotation around a common center of the galaxies; and so on. For Kant, in other words, the Newtonian procedure for determining a privileged frame of reference is necessarily nonterminating: it aims ultimately at the "center of gravity of all matter" (p. 131), and this point lies forever beyond our reach. In place of Newton's absolute space we are therefore left with no object at all, but only with a procedure for determining better and better *approximations* to a privileged frame of reference: from a modern point of view, a privileged inertial state. This is why Kant calls absolute space an Idea of Reason in the final chapter, or Phenomenology, of the *Metaphysical Foundations;* for absolute space can be thought of only as the ideal end point toward which the Newtonian procedure for determining the center of mass is converging, as it were.

By the same token, however, Kant holds that all of our inquiries into nature are directed by Ideas of Reason which provide regulative ideals for the methodological practice of science. Such regulative ideals must ultimately be unified in a single Idea of Reason (which Kant naturally calls the Ideal of Pure Reason): namely, the Idea of a Wise Author of the world. That is, for the purpose of guiding our inquiry into nature, we declare "that the things in the world must be viewed *as if* they received their existence from a highest intelligence."[19] Although Kant insists that this last conception is only an

Idea, and we cannot therefore assume an object corresponding to it, this conception of a Wise Author is still not even an *Idea* of Divinity. For the latter we require the additional attributes of omniscience, omnipotence, omnipresence, and omnibenevolence, which can only flow from the conception of the Highest Good to be produced in the world, a conception which can itself only be based on our notion of an ideal *moral* order of things; and so, in order to complete the Idea of Divinity, we therefore require the construction of the moral law by pure *practical* reason.[20] In other words, only the conception of an ideal moral commonwealth can finally complete the regulative function of reason; and this, for Kant, is what secures the *unity* of practical and theoretical reason.

Kant sums up this last train of thought in a striking footnote to the General Observation to Book III of *Religion within the Limits of Reason Alone* (1793):

> Similarly, the *cause* of the universal gravity of all matter in the world is unknown to us, so much so, indeed, that we can even see that we shall never know it: for the very concept of matter presupposes it as a primary moving force unconditionally inhering in it. Yet gravity is no mystery but can be made public to all, for its *law* is sufficiently known. When Newton represents it as similar to divine omnipresence (*omnipraesentia phaenomenon*), this is not an attempt to explain it (for the existence of God in space involves a contradiction), but a sublime analogy which has regard solely to the union of corporeal beings into a world-whole, in so far as one bases this on an incorporeal cause. The same result would follow upon an attempt to comprehend the self-sufficient principle of the union of rational beings in the world into an ethical state, and to explain this in terms of that principle. All we know is the duty which draws us towards such a union; the possibility of the achievement held in view when we obey that duty lies wholly beyond the limits of our insight.[21]

Of course Kant is not here adopting a Newtonian conception of space, gravity, and divinity but, once again, fundamentally transforming such a conception. Instead of absolute space conceived of as a fixed arena within which all states of motion are automatically well defined, we have a nonterminating but convergent methodological procedure for approximating a privileged frame of reference. Instead of gravity conceived of as an activating power which requires the intervention of a medium—most likely an immaterial medium—we have a mathematical law of motion whose two properties of immediacy and universality are necessarily presupposed in this methodological procedure. Instead of a Living God who fills all space with real omnipresence, we have a *focus imaginarius* toward which all our activities, both practical and theoretical, ideally converge. As I have emphasized, however, we should never forget that this fundamental Kantian transformation is itself effected—at least in the case of the first two elements, space and gravity—only on the basis of Kant's extraordinary insight into the conceptual foundations of Newton's divine *Principia*.

Notes

1. I. Newton, *Mathematical Principles of Natural Philosophy*, trans. A. Motte, rev. F. Cajori. Berkeley, 1934.
2. I. Kant, *Metaphysical Foundations of Natural Science*, trans. J. Ellington. Indianapolis, 1970. I have emended Ellington's translation in the quotations that follow. References to this work are given parenthetically in the text.
3. I. Kant, *Critique of Pure Reason* (first edition 1781, second edition 1787), trans. N. Kemp Smith. New York, 1929.
4. *Göttingen Anzeigen von gelehrten Sachen*, January 19, 1782, pp. 40 ff. Kant explicitly responds to this review in the Appendix to his *Prolegomena to Any Future Metaphysics* (1783), trans. L. Beck. Indianapolis, 1950.
5. Kant's quotation (in Latin) occurs on p. 66 of *op. cit.*; see I. Newton, *Opticks*, based on the 4th ed. New York, 1952, following p. cxxii.
6. *Opticks*, p. 400.
7. *Mathematical Principles*, pp. 415–417.
8. See the footnote to the preface to the second edition of the first *Critique* at Bxxii.
9. This notion of an "acceleration-field" is introduced in Howard Stein's seminal paper, "Newtonian Space-Time," *Texas Quarterly* (1967): 174-200, to which I am greatly indebted throughout my discussion of the argument of Book III of *Principia*.
10. For more details here see M. Friedman, "The Metaphysical Foundations of Newtonian Science," in R. Butts, ed., *Kant's Philosophy of Physical Science*. Dordrecht, 1986, pp. 25–50.
11. J. Edleston, ed., *Correspondence of Sir Isaac Newton and Professor Cotes*. London, 1850, pp. 152 ff. For discussion see A. Koyré, "Attraction, Newton, and Cotes," in *Newtonian Studies*. Chicago, 1968, pp. 273–282.
12. See C. Wilson, "The Great Inequality of Jupiter and Saturn: from Kepler to Laplace," *Archive for the History of Exact Sciences* 33 (1985): pp. 67–69.
13. For the importance of light in establishing the perceptible community of the heavenly bodies, see the first *Critique* at A213-214/B260-261.
14. I. Cohen, ed., *Isaac Newton's Papers & Letters on Natural Philosophy*. Cambridge, Mass., 1978, p. 302.
15. Kant, in his *Universal Natural History and Theory of the Heavens* (1755), trans. W. Hastie. Glasgow, 1900, also explicitly opposes this conviction.
16. I. Cohen, *op. cit.*, pp. 302–303.
17. Parts II and III of Bentley's *Confutation* are reprinted in *ibid.*, pp. 313–394.
18. *Opticks*, p. 403.
19. *Critique of Pure Reason*, A670–671/B698–699.
20. See I. Kant, *Critique of Judgment*, trans. J. Meredith. Oxford, 1952, §86.
21. I. Kant, *Religion within the Limits of Reason Alone*, trans. T. Greene and H. Hudson. New York, 1960, pp. 129–130; translation slightly emended.

Chapter 11

The "Essential Properties" of Matter, Space, and Time: Comments on Michael Friedman

Robert DiSalle

Michael Friedman has reconstructed Kant's struggle with two of the most important philosophical problems posed by Newton's physics: how to reconcile the structure of space and time with the Galilean relativity of Newton's mechanics, and how to give a physical interpretation to the gravitational field. Both of these are challenges to what Friedman calls "the older tradition in metaphysics," and it is tempting to conclude from his analysis that Kant's understanding of these problems, and especially of the interconnections between them, constitutes an important constructive critique of Newton's. I would like to explain why we should resist that temptation by considering three questions.

First, what is mistaken about Newton's conception of absolute space, and did Kant supply a useful correction? Recall that Newton discusses space and time in the Scholium to the Definitions in the *Principia*, before he introduces the laws of motion: evidently he assumes that since the laws involve spatiotemporal quantities like momentum and acceleration, he must specify the spatiotemporal background against which the laws are to be understood; as he points out, these laws will not hold in just any relative space, so an absolute space must be assumed if they are to be universal laws. What Newton did not grasp, we now see, is that his laws of motion require absolute time but *not* absolute space: they actually entail a weaker space-time structure, which Professor Friedman has called "Galilean space-time." What Newton should have said, in the language available to him, is that his laws posit at least one space in which motions have the three characteristics described by those laws, and that any space in uniform irrotational motion relative to this one is dynamically equivalent to it.

I apologize for rehearsing a rather familiar point, but I wanted to restate the second part of my question more precisely: Does Kant's rejection of absolute space represent progress from the structure postulated by Newton toward the weaker structure postulated by Newton's laws? As Friedman shows, Kant makes absolute space a "mere idea," but obviously this is a separate issue from that of characterizing the necessary structure. On the one hand, one might say that Galilean space-time (or any relativistic space-time) cannot be an object of experience and hence is a mere idea; on the other hand, absolute space taken as a regulative idea might, no less than Newton's real absolute space, turn out to introduce superfluous structure into physics. I believe that the second possibility is just what we find realized in Kant's *Metaphysical Foundations of Natural Science*.[1] In the Phoronomy he defines absolute space as "that in which all motion must ultimately be thought" (p. 480), but he cautions that "to make this an actual thing means to mistake the logical universality of any space, with which I can compare each empirical space as being included in it, for a physical universality of actual extension" (p. 842). But in the Phenomenology he names "the concept of motion in absolute (immovable) space" as one of the concepts "whose employment in universal natural science is unavoidable" (p. 559). In order to understand his position, and to answer my question, we need to focus on that "employment."

In one important respect Kant's use of absolute space seems to me precisely parallel to Newton's. Kant writes, "Absolute motion, that is, such as is thought without any relation of one matter to another, is impossible"; but, he adds, "that is just why no concept of rest or motion in relative space is possible that is valid for all appearances" (p. 559). "Absolute space. . . is necessary as an idea that is to serve as a rule for considering all motion within it as merely relative, and all motion and rest must be reduced to absolute space, if appearance itself is to be transformed into a determinate concept of experience ([a concept] that unites all appearances)" (p. 560). This appears to be a version (though a version transformed, obviously, by Kant's epistemology) of Newton's argument that absolute space is necessary because no empirical space can be considered as at rest—because no standard of motion provided by an observed body would be "valid for all appearances." This implies, for Newton, that absolute space is necessary in spite of his fifth Corollary, not indeed because a certain space can be physically distinguished from all others, but because the class of equivalent relative spaces named in the corollary is defined *only* with respect to the encompassing space, along with absolute time, with respect to both of which they move "uniformly in a right line without any circular motion."

Thus both Newton and Kant place absolute space at the foundations of physics as the structure that makes possible the treatment of relative motions in relative spaces (differing, again, on the metaphysical status of the structure). It was always clear to Newton that the accompanying notions of absolute rest and velocity play no further role in dynamics; that is a crucial part of the content of his remarkable Corollary V. But Kant's *Metaphysical Foundations* expresses no clear grasp of that content: in place of the very precise Corollary and its detailed proof from the laws of motion, Kant offers the comparatively vague Proposition I of the Phenomenology: "The rectilinear motion of a matter in regard to an empirical space, as distinguished from the contrary motion of the space, is a merely possible predicate" (p. 555). Now, Kant's next proposition indicates that he does not extend this "equivalence of hypotheses" to circular motions; but what about nonuniform motions? Kant consistently speaks of the equivalence of states of *rectilinear* motion, but, as Newton's Corollary points out, the class of privileged frames has to be defined with respect to *time* as well as to space—they lack acceleration as well as rotation—and Kant gives no clear statement of this point. (It is worth noting that the argument for Newton's Corollary V depends crucially on his *second* law of motion, which Kant does not include among the fundamental a priori laws of mechanics.)

Possibly Kant did understand "Galilean" relativity precisely as Newton explained it, in which case I have merely pointed to a defect of Kant's presentation; but if we ask whether he, like Newton, always respected this relativity principle in dynamical reasoning, we are again led into difficulties. In the Mechanics he remarks that the equivalence of hypotheses mentioned in the Phoronomy does not hold for interactions among bodies: "It is no longer the same whether I attribute an opposite motion to one of these bodies or to the space," for here we consider not just the velocity, which is relative, but also "the quantity of substance (as moving cause)"; and so "it is no longer optional, but *necessary* to assume each [of two bodies in an impact] to be moved, and indeed with equal quantities of motion in opposite directions" (p. 547). (This last clause is the basis for Kant's claim that the opposing motions in an impact "destroy one another in absolute space.") This argument relies on the notion of moving force or impulse, and it leads to a further claim that only Newton's first law of motion should be called the law of inertia: that law decribes the *inactivity* of matter, and this inactivity alone "is not the cause of resistance"; "nothing can resist a motion but the opposite motion of another, never the other's rest" (p. 551). Howard Stein has emphasized that Newton took inertia to be defined by all three of his laws of motion, and Kant's arguments show that the difference between the two is more than verbal: Kant is apparently trying to say that the amount of relative motion is the important fact in a collision, but his distinction of impulse from inertia forces him to appeal to absolute space. Compare, by contrast, Newton's lucid account in his definition of the *vis inertiae:*

> A body exerts this force only when another force, impressed upon it, endeavors to change its condition; and the exercise of this force may be considered both as resistance and impulse; it is resistance, in so far as the body, for maintaining its present state, withstands the force impressed; it is impulse, in so far as the body, by not easily giving way to the impressed force of another, endeavors to change the state of that other. Resistance is usually ascribed to bodies at rest, and impulse to those in motion; but motion and rest, as commonly conceived, are only relatively distinguished; nor are those bodies always truly at rest, which are commonly taken to be so.[2]

The comparison suggests that Kant is granting an absolute status to something that, from Newton's point of view, "depends on one's frame of reference."

All of these considerations seem to me to cast doubt on the claim that Kant has offered a useful critique of Newton's absolute space; my next question concerns the critique of Newton's view of gravity: how should we interpret Kant's claim that gravity is essential to matter, and is it a plausible claim? If we understand it in a narrow sense, meaning only that gravitation must be a universal and immediate action at a distance, then Kant's argument about the third law is indeed plausible; Howard Stein has suggested that arguments of this sort lay behind Newton's increasing pessimism concerning the possibility of an "aetherial mechanism" for gravity and his tendency (expressed in Query 31 to the *Opticks*) toward the view that the law of gravity may simply be among the laws that God imposed at the Creation.[3] (This view should be distinguished, by the way, from the view that God is directly involved in all gravitational accelerations—a view that Professor Friedman seems to attribute to Newton. In Query 31 Newton names gravity among the "general laws of nature, by which the things themselves are formed."[4] Once God has decreed them, however, such laws seem to constrain him just as the principle of sufficient reason constrained the God of Leibniz: the gravitational interactions of planets and comets will disturb the system of the world so that God will be called on to reform it; similarly, God had to place the stars at immense distances from one another "lest the systems of the fixed stars should, by their gravity, fall into one another mutually."[5]

But we want to understand why gravity is "necessary to the possibility of the concept of matter." Here, if we accept Friedman's interpretation—that gravity is essential to the concept of matter as the movable in space—then the claim does not seem plausible at all. The argument turns on the use of universal gravitation to estimate masses; but, as Kant says in Proposition I of the Mechanics, "The quantity of matter can be estimated in comparison with every other matter only by the quantity of motion at a given velocity" (p. 537), and so any interaction in which momentum figures will enable us to estimate masses—for example, particle collisions. Moreover, Kant appears to recognize this fact when he says that "the quantity of substance in a matter must be estimated mechanically, that is, by the quantity of the proper motion of the matter, and not dynamically, by the quantity of its original moving forces" (p. 541). If gravity can "nevertheless" be used to estimate masses, this is because such an estimation is *indirectly* mechanical (p. 541); the gravitational measure of mass is therefore distinctly secondary in Kant's analysis. Thus, that gravity is essential to matter as the measure of mass seems implausible not only from my point of view but from Kant's also.

My last question concerns the interconnection between gravity and absolute space: How is the "original force of attraction" relevant to the construction of a privileged frame of reference? If what I have already said is correct, then gravity cannot be *essential* to such a construction; collision experiments will suffice to tell us whether any given frame is inertial. It is true that only gravity can tell us the masses of the planets, because that is the only interaction among them that involves mass, but this is not a limitation *in principle*: it reflects the circumstances that (for example) any projectile we are capable of hurling at other planets will strike with a negligible momentum. Thus gravity was indeed essential for Newton to treat the planetary system within an inertial frame, but, independently of gravity, purely mechanical experiments will suffice to ground the concept of mass, the laws of motion, and the notion of a privileged frame. Indeed, it was in connection with such experiments that all three originally evolved. If Kant did hold the opinion that Friedman attributes to him, then we must admit that Kant was mistaken.

But I have already given reasons to suspect that Kant may not have held such a view. A more interesting answer to the question, then, must address the respective metaphysical assumptions of Newton and Kant. For Newton, what is essential to matter, or to "brute inanimate matter"? His three "laws of inertia" suffice to define "matter as the movable in space," for they explain the natural state of motion of bodies, the manner in which changes of motion arise from interactions, and how masses and relative velocities enter into interactions; as Newton recognized, moreover, we cannot claim that matter has the properties described by his laws without making reference to a certain space-time structure, and from the background structure and the laws together he derives the class of inertial frames. Now, why are these "passive principles" essential, rather than the active forces postulated by Kant? They are essential not from an epistemological, nor even from a metaphysical, but from a methodological point of view: they are the foundation for Newton's program for *discovering* what active principles are at work in the world; these principles, derived from terrestrial experiments, make it possible to bring the celestial bodies under the concept of matter, and in the process to show that acceleration fields surround them. This is why Book I of the *Principia* commands as much philosophical interest as Book III: it gives a rigorous general account of how deductions from the passive principles guide us in exploring the active forces of nature. The application of this program may lead us to the conclusion that a certain force is "essential to matter," but such a conclusion will not represent an epistemological constraint; rather, it will indicate that we have learned something surprising about matter.

Kant, in contrast, presents the active forces first and seems to place them on a par with the two passive principles he postulates (Newton's first and third laws; and, as we saw, for Kant only the first is truly passive). On Professor Friedman's reading, this presentation is part of an "operational" approach to space-time structure which aims to arrive at a privileged frame with the help of physical principles (and with the particular help of the force of gravity), again an approach made necessary by the fact that absolute space cannot be an object of experience. In my view, however, it is difficult to reconcile such an approach with Kant's view of what is essential to matter. For he claims that inertia is among the essential properties, and this claim makes no sense unless we presuppose some of the structure of Newtonian space-time, for Kant's explanation of inertia appeals to a "natural state" of motion. Similarly, Kant introduces the "original attractive force" before explaining the "law of inertia," but this notion cannot be used to *define* the *concept* of a privileged frame; as Newton realized (and as his Book I effectively shows), the very notion of an "acceleration-field" is meaningful only against the background of the passive laws of motion and the space-time structure on which they are founded. Operationally, Newton relied on gravitational attraction in arguing that his conception of motion actually applies to the system of the world, but no such force is required to make that conception coherent, either as an abstract structure or as a set of practical rules. Finally, by not recognizing the connection between laws of motion and space-time structure, Kant places himself, metaphysically speaking, in a questionable position: if inertia and acceleration fields are essential to matter, how can space be a mere idea?

I would like to add a remark concerning the relativity of motion. To borrow Friedman's phrase, on this point Kant seems to belong to an older tradition in metaphysics, which Newton tried to transform. According to the older tradition (which obviously did not end with Kant), the relativity of motion is a fundamental philosophical truth with which physics has to come to grips; this is why Kant takes the equivalence of rectilinear motions as a fundamental truth. Newton starts with a slightly different premise: that the world presents us with relative motions which physics has to try to understand. The performance of this task begins with laws of motion; an idea that there is a class of equivalent relative spaces—a "theory of relativity"—is for Newton an aspect of these laws rather than a prior epistemological or metaphysical commitment. Thus, if the claim that "motion is relative" expresses a philosophical insight rather than simply the predicament in which physics begins, it is because the precise nature and extent of the "equivalence of hypotheses" is grounded in physical theory and warranted by the application of the theory to experience. It seems to me that philosophers' successful efforts during the last few decades to understand the consequences of Einstein's theories have led us back to this insight of Newton's.

Notes

1. Immanuel Kant, *Metaphysische Anfangsgründe der Naturwissenschaft* (1786), from Kant's *Gesammelte Schriften*, Akademie Ausgabe, vol. 4 (Berlin: Reimer, 1911). References to the work will be given in the text as page numbers from this edition.
2. From Definition 3 of Isaac Newton, *The Mathematical Principles of Natural Philosophy*, trans. Andrew Motte (London, 1728; reprint, New York: Philosophical Library, 1964), pp. 13–14.
3. Howard Stein, "Newtonian Space-Time," *Texas Quarterly* (August 1967), 174–200.
4. Isaac Newton, *Opticks* (London, 1730; reprint, New York: Dover, 1952), p. 401.
5. From the General Scholium, in Newton, *The Mathematical Principles of Natural Philosophy*, p. 444.

Chapter 12

Ethical Implications of Newtonian Science

Errol Harris

i

The publication of Newton's *Principia* three centuries ago brought to fruition the Copernican revolution, by which the conceptual scheme was established involving a world view contrasting sharply with its predecessor (the one typical of the ancients and the medievals).

As is well known, the world, for the ancients, was a living being with an all-pervasive soul, of which the souls of the gods, of humans, and of animals were individual centers. In such a world there was no breach between humankind and the rest of nature, nor any great gulf fixed between the gods and the rest of the world. Thales of Miletus is reputed to have said that all things are full of gods, and the soul (whether of men or beasts) was held to be simply the most refined form of whatever was taken to be the primal substance, of which all things were made. The moral life of men was seen in terms of religion—a duty toward the gods, conceived as natural deities—and of virtue, the duty to oneself and one's fellow citizens, required by nature itself. For man was by nature a rational and a social animal, and morality was what the city enjoined and what reason prescribed.

In the Middle Ages this world view was modified only so far as the Judeo-Christian beliefs required. The world was God's creation, and what happened in it was God's act. The law of nature was God's law, for he had created a rational and moral order, giving mankind reason and free will. The nature of things was thus both reasonable and moral, and natural law established a rule for moral conduct as well as an order of physical motion.

The Newtonian world, on the other hand, was a vast machine, devoid of soul and mind—an artifact devised by the mind of its creator, the Supreme Architect, and an object of knowledge both to his mind and to those of human beings. But neither of these found any place within the celestial system itself. God had made and presided over the heavens and the earth, and man observed and could discover the mechanisms of the celestial motions.

This contrast between the seventeenth-century outlook and that of preceding ages becomes clear from Newton's assertion in the General Scholium at the end of Book III of the *Principia*, where he writes: "This Being [God] governs all things, not as the soul of the world, but as Lord over all; and on account of his dominion he is wont to be called Lord God."[1]

Descartes had expressed much the same conception in the fifth part of his *Discours de la methode*. There he says:

> It is certain, . . . that the action by which he [God] conserves it now is the same as that by which he created it [i.e., the world]; so that even though he did not at the beginning give it any other form than of chaos, provided that he had established the laws of nature and lent it his preserving action to allow it to act as it does customarily, one can believe, without discrediting the miracle of creation, that in this way alone, all things which are purely material could in time have made themselves such as we see them today.[2]

From these quotations it is apparent that the world is conceived as God's artifact, and, the laws of nature once established—and these, for Newton, are the laws of motion formulated in the *Principia*—the machine will continue to run of its own accord, in obedience to them.

In this Weltanschauung the world consists entirely of material bodies in motion. In the words of another seventeenth-century thinker, Thomas Hobbes:

> The World (I mean not the Earth onely, . . . but the *Universe*, that is, the whole masse of things that are) is Corporeall, that is to say, Body; and hath the dimensions of Magnitude, namely, Length, Bredth, and Depth: also every part of Body, is likewise Body, and hath the like dimensions; and consequently every part of the Universe, is Body; and that which is not Body, is no part of the Universe: And because the Universe is All, that which is no part of it, is *Nothing*; and consequently *no where*.[3]

This, of course, includes the human body, which is likewise a machine. As Hobbes puts it (in the introduction to *Leviathan*):

> For seeing life is but a motion of Limbs, the beginning whereof is in some principall part within; why may we not say, that all *Automata* (Engines that move themselves by springs and wheeles as doth a watch) have an artificial life? For what is the *Heart*, but a *Spring;* and the *Nerves*, but so many *Strings;* and the *Joynts*, but so many *Wheeles*, giving motion to the whole Body, such as was intended by the Artificer?[4]

It was generally agreed that in God, knowing, creating, and sustaining were all one single act. But for human kind, knowing and doing were distinct actions. The world appears to the human mind, and is not made by it. For us the world is phenomenal, and Newton tells us in the Scholium, just quoted, that from the phenomena "particular propositions" may be inferred. Among these "particular propositions" are the laws of motion. But how, it must be asked, can anything appear to a body that is no more than a machine? How in this material universe is the mind to be accommodated? The world, surely, will appear through the senses, and, again, Hobbes tells us that "qualities called *Sensible*, are in the object that causeth them, but so may several motions of the matter, by which it presseth our organs diversely." "Neither in us that are pressed, are they any thing else, but diverse motions; (for motion produceth nothing but motion.) But their appearance to us is Fancy, the same waking that dreaming."[5]

What Hobbes failed to explain, or even to consider, was what "fancy" could be in, or how it could appear to, a body consisting of nothing but springs, strings, and wheels. Small wonder that Descartes regarded cognitive consciousness as a different substance, altogether separate and distinct from extension, communicating with human bodies mysteriously through the pineal gland by virtue of divine intervention.

Thus the effect of Newtonianism on epistemology was that, because the object of science was external to, and exclusive of, mind, it could (apparently) become accessible to knowledge only through its imping on the human body—that is, through the senses. Admittedly nobody could explain how physical impulses were, or could be, translated into mental contents, how motions and pressures could be converted into "fancy." Yet that we are apprised of surrounding bodies by sensation seemed obvious; so sense perception took precedence over reflection and thought. Reason was reduced to the combination, separation, and comparison of ideas derived from sense, and logic took the form of a sort of verbal arithmetic: as Hobbes put it, "Reason . . . is nothing but *Reckoning* . . . the consequences of general names." As a result, reason could function only instrumentally, and this had a concomitant effect on ethics. In practical affairs, reason might calculate and compare the means to ends sought from desire but could never function constructively to define objective standards of value.

The result is a sharp division between fact and value. The latter is purely subjective, originating in human desire, a merely human sentiment, expressed, no doubt, as if it applied factually to external things, although in actuality it was only an expression of feeling.

Hume (and others before him) had traced moral judgment to a moral sense, to which reason, "the slave of the passions," was merely ancillary.

But for the scientist the objectivity of the world was its mechanical materiality, devoid of, and external to, mind. It could be observed, by whatever means, only from without; and if it were to be observed accurately, its stark externality must be maintained. Nothing subjective must be imported into it by the observing mind. Even secondary qualities were suspect (as mind dependent), and only the primary, measurable and mathematizable qualities could be scientifically known. Evaluations, other than numerical, dependent as they were on feeling, sentiment, emotion and desire, all subjective elements, if they were pronounced of bodies at all, must be considered as not even secondary but rather as tertiary qualities. They were to be altogether excluded from scientific description, which had to be entirely value free.

Value theory, ethics, and aesthetics, therefore, initially did not fall within the ambit of Newtonian science, although, by long tradition they had been and remained philosophical disciplines. But under the influence of the new science they were confined to the sphere of human subjectivity, human aspiration, desire, and satisfaction. In consequence the ethical speculation of the succeeding three centuries has been confined, for the most part, to two main types, utilitarian and deontic. Moral goodness has been conceived purely in terms of the satisfaction of human desire and the attainment of happiness; and moral obligation as owed solely to human persons. The physical and biological world was not regarded as morally relevant, except as providing means to human ends. Nature might and should be exploited, and human needs took precedence over the welfare, or survival, of animals, which Descartes had regarded as automata, much as Hobbes had described the human body. So we find Hume, Bentham, and the Mills identifying moral goodness with pleasure and happiness, virtue with benevolence and rational self-interest; and Kant defining moral obligation in terms of a universal law to treat all humanity as an end in itself, as opposed to the simply natural, which may legitimately be treated as a mere means.

The medieval doctrine of natural law, postulating an objective moral order continuous with, and indeed integral to, the order of nature, came to be castigated as fallacious. So modern ethical writers have become obsessed with, and have discussed *ad nauseam*, "the naturalistic fallacy": that of attributing value to things on account of their natural properties, and of attempting to derive moral obligation from existing (or possible) states of affairs. "Ought," we have been assured since Hume, cannot be deduced from "is," and evaluative statements are never legitimately indicative but are always open or disguised imperatives or optatives. Thus normative ethics becomes a noncognitive discipline, not primarily concerned with the objective nature of things but only with human feelings and desires and their consequent exhortations and prescriptions.

Newtonian science, however, was so spectacularly successful in the realm of physics that its methods were progressively applied, first to chemistry, and then, by a natural extension, to biology, and finally to psychology and the social sciences. But in order for this to occur, the subject matters of these sciences had to be viewed in the same way as physical phenomena: as objective entities and events, following general laws, to be observed dispassionately from without. Human feelings and desires, when they become the objects of empirical psychology, can be so observed, and so may human habits of conduct. Human action is then reduced to behavior, publicly observable, like that of other animals, and with it preferences and evaluative behavior become phenomena to be disinterestedly described and recorded. For such a scientific approach, the question whether moral judgments are legitimate or valid does not arise. They are no more than observable events the description of which must be value free.

Treated in this way, human feeling and desire are seen to be subjective to individuals, and variable from one person to another; and so likewise are moral evaluations. Moreover, as human beings are social animals, their behavior toward one another is habituated by social tradition and regulated by socially conditioned rules. These become the objects of anthropology and sociology, and these sciences discover that rules of conduct and codes of morals are peculiar to the cultures within which they prevail. They vary from one such culture to another, and do so to such a degree that often what is approved in one society is disapproved in another, and what is condemned among one people is acclaimed among others. For similar reasons this can occur even within a particular culture where there are subcultures, differing social conditions, and divergent preferences.

A similar conclusion emerges from the study of history. Moral codes and beliefs change with time, and those of one age differ from those of another, so that present standards cannot rightly be applied to past actions, which, to be properly understood, must be seen in the light of the criteria prevailing at the time.

Consequently, standards of value, moral and other, have been universally found to be relative—to the historical period, to the particular culture, to the special subculture, and even to individuals—and no other assertion can be universally stated about them. This is the general finding of social anthropologists, and the prevailing view with respect to morality.

ii

To this day ethical theory has remained under the spell of the scientific outlook initiated in the seventeenth century, and it is still conditioned by the Cartesian dichotomy, the rift between nature and mind. So G. E. Moore distinguishes, in *Principia Ethica*, between natural and nonnatural qualities in his attempt to discover the nature of goodness.[6] Fact remains on one side of the divide and value on the other; as Bertrand Russell maintained, "A judgement of fact is capable of a property called 'truth,' which it has or does not have quite independently of what any one may think about it . . . But . . . I see no property, analogous to 'truth,' that belongs or does not belong to an ethical judgement. This, it must be admitted, puts ethics in a different category from science."[7] Ethics has continued to treat morality from a subjectivist point of view and valuation as a subjective propensity. In so doing it is not, of course, wholly misguided, because there is no evaluation without self-awareness, reflection, and deliberate choice. Nevertheless, the question concerning objective standards remains open. At all events, the persistent trend of modern ethics has been subjectivist in all its predominant forms.

Empiricist ethics has constantly been hedonistic and utilitarian. Even ideal utilitarianism has tacitly assumed that the criterion of moral goodness is ultimately pleasure; and pleasure, whether of the self or of the greatest number, is a subjective feeling, and when made the ground of approval that too is taken to be no more than a feeling. The extreme form of this doctrine is the emotive theory of ethics of such writers as A. J. Ayer and C. L. Stevenson, declaring that moral statements are unverifiable and can therefore never be assertions of belief (which must be either true or false) but can only be expressions of feeling, of approval or disapproval, or expletives designed to persuade or cajole others into adopting attitudes similar to one's own toward certain types of actions and those who commit them. What some approve, however, others disapprove, and what some find pleasant others do not—*De gustibus non disputandum*—so this type of ethic cannot but be relativistic.

But there is no need to go to such extremes. The prevalence of utilitarianism, in one form or another, in contemporary ethical theory can be seen by reading a book like Kurt Baier's *The Moral Point of View*,[8] in which it is not difficult to detect below the surface the presumption that "enjoyment" is the primary "reason" for action, while the explanation of the obligation to self-restraint and self-regulation and the superiority of "moral reasons" is that it is the sole means of avoiding the unmitigated discord and conflict that Hobbes describes as "the State of Nature." Thus the appeal to pleasure (or the avoidance of pain) remains the source of the preference. The rules to be observed for this purpose, according to Baier, are those inculcated in youth by a society and enforced by its established authority; hence the moral code is relative to the social group.

Deontic theories (like those of W. D. Ross and H. A. Prichard) declare moral duty to be absolute and its content to be self-evident; we recognize intuitively what is right and obligatory: for example, the duty to fulfill promises. But it has ultimately to be confessed that intuitions differ from person to person, and what is felt to be obligatory from one society to another, so relativism has not been avoided. Much the same is true of the existentialist doctrine that choice (which is free) is prior to moral judgment. For freedom of choice leaves the decision open to the individual, so that the commitment assumed will vary with the agent.

Even moralists who are at pains to refute subjectivism find difficulty in identifying the objective standard, which, they maintain, is always presupposed in moral judgment and ethical language. Yet it is generally admitted that the performance of duty is conditional upon beliefs about circumstances, about the probable consequences of the action, and (most important) about moral principles, and such beliefs are necessarily subjective; they are liable to vary between societies and even among their individual members.

It may be objected that many of these theories have no direct relation to Newtonian science, and certainly it may be conceded that this is the case and that they are the direct product of reflection upon common moral experience. Nevertheless that experience itself has been affected by the common view of the physical world as devoid of mind and, except as providing means to human ends, also of value—as not being an end in itself. The belief is also widespread that morality is inevitably subjective and that no moral value inheres intrinsically in the natural world. The crucial question then arises as to whether, or to what extent, human action itself is purely natural or is independent of natural causes. In the running dispute over this issue the Newtonian conception of causality is central and decisive. If human action, along with the physical world, is subject to Laplacian determinism, moral praise or blame become otiose; and then the springs of human choice must somehow be found outside and beyond the world of nature. Dispute on this matter has dogged the heels of modern ethics throughout its history and has not yet ceased to trouble it.

The Cartesian dualism, born of the seventeenth-century exclusion of mind from the physical world, still infected Kant's attempt to resolve this problem. He postulated a double causation, one empirical and necessary, the other noumenal and free. Human action, so far as it is derivative from the spontaneous activity of the transcendental ego, which cannot be brought under the categories, and is therefore noumenal, is free; but as a natural phenomenon conditioned by sense and desire, it is determined by natural laws. But Kant could not provide the rational will, the sole object of moral judgment (nothing but the good will being good without qualification), with any concrete empirical content, save that its action should be for its own sake; and the desires conditioning the behavior of the empirical self remained incorrigibly vicious and phenomenally caused. Empirically, therefore, the moral law requiring universality remained either empty or impotent, or both, and the determinants of the natural self both subjective and unfree. Moral objectivity, in consequence, was inapplicable to phenomenal behavior, and empirical necessity was imposed only by the categories of the understanding determining facts but incompetent to prescribe moral laws and duties. The introjection of noumenal causation into the phenomenal world, meanwhile, was futile without empirical content.

In short, objective criteria were provided exclusively by empirical science, while moral values remained dependent solely on human judgment and human desire. The former, if it is to be free, cannot be subject to those objective criteria, and the latter are incurably subjective. So the door is left open to the relativism sponsored by sociology and social anthropology, conditioned, as they are, by the Newtonian outlook.

My thesis gains impressive support from Alasdair MacIntyre's admirable analysis of the modern ethos and its philosophical counterparts in his book *After Virtue*.[9] The support is both direct and indirect: direct because his assessment and description of the development of ethical theory since the seventeenth century, though more detailed and penetrating, is very close to my own; indirect because his position and argument are historicist, and his conclusion concerning the conditions necessary for the reestablishment of the moral tradition still leaves the way open for relativism.

MacIntyre argues, and who could disagree, that contemporary moral theory is emotivist, not only in the hands of A. J. Ayer and C. L. Stevenson but also, by implication, for existentialists like Kierkegaard, Nietzsche, and Sartre, and that this emotivism is the product of the successive failures of such philosophers as Hume and Diderot, with their utilitarian successors, on the one hand, and Kant, followed by the deontologists (Prichard and Ross), on the other, to supply any rational vindication for conventional morality (as handed down from the Greeks and early Christianity), after the repudiation by the Enlightenment of Aristotelian teleology and medieval authoritarianism. What I have attributed to the influence of the presuppositions of Newtonian science MacIntyre attributes to the Enlightenment's claim to have liberated ethics from the leading strings of theology and the teleological biology and metaphysics of Aristotle, establishing the autonomy of the moral self, free from the authority of the church. He omits to note that this Enlightenment claim is the direct consequence of the Copernican revolution and its consolidation by Galileo and Newton, with the resulting abandonment of Aristotelian physics and metaphysics. This produced the new mechanistic world view, with the consequences I have outlined.

MacIntyre maintains that modern ethics in large measure founders on its separation of fact from value, which, he shows, is not a timeless logical truth but simply the result of a change in the use of evaluative expressions, made during the late seventeenth and early eighteenth centuries, such that what at that time ranked as factual premises could no longer entail what now passed for evaluative and moral conclusions. MacIntyre thus reinforces my brief remarks on the subjectivization of moral judgments, due to the divorce of scientific fact from value.

Still more interestingly, MacIntyre demonstrates that the common moral outlook of the present day and the personalities typical of modern moral attitudes presuppose an absence of any belief in objective standards of value. Manipulation substitutes for respect of persons; aims and commitments have become partisan and incommensurable, and no ultimate end is envisaged for human endeavor, apart from unspecified "success." Consequently, the current use of moral terms and apparent deference to rational precepts become largely a front (recognized as such by theorists) for propaganda in favor of personal preferences and arbitrary commitment to particular (partisan) social ends. In short, both in practice and in theory the present age is abandoned to ethical scepticism, subjectivism, and relativism.

All this, MacIntyre asserts, results from the presumption by theorists, Nietzsche in particular, that the moral tradition, inherited from the Greeks and Christianity, has broken down, and that its surviving relics have not been, and cannot be, given any rational foundation. But MacIntyre believes that this breakdown of the tradition cannot, in the final issue, be substantiated, and that with some necessary and salutary modifications it can be given a sound basis in a certain kind of communal practice. So Nietzschean cynicism turns out after all to be just another facet of the very moral culture it took itself to be criticizing. The revival and reestablishment, however, of a kind of neo-Aristotelian moral tradition must be based on community action aimed at common purposes, which will give narrative coherence to the lives of the people involved; and, although this is, I should agree, a sound foundation for the rebirth of morality, unless the community envisaged were to encompass the whole of humanity, so that the tradition became universal, each separate society would engender its own peculiar code of morals. Thus MacIntyre's conclusion, while it is not itself necessarily relativistic, does still leave scope to the anthropologist to argue that morals are inevitably relative to the particular culture in which they are practiced.

iii

The difficulty with relativism, of course, is that it undermines the very basis of morality and leads to ethical scepticism. The pursuit of ethical speculation issues from a practical pretext. The question arises for every thinking person at some time: What is the best way for me to conduct my life? Whether one has been brought up in the habit of obeying the prescripts of some conventional morality or left to one's own devices, conflicts of duties (in the first case) or conflicts of desires (in the second) will prompt one to raise such a question. "Why should I be moral?" was the form in which it was asked by F. H. Bradley in the second of his *Ethical Studies*.[10] But the question implies a preference for one course of action over others, and so requires a decision as to the criterion by which the preference is made. If the criterion turns out to be purely personal and subjective, it is no more than caprice, and the distinction fundamental to morality between inclination and duty disappears. If it is relative to a particular community, it remains arbitrary until it can be related to an objective criterion universal to all. Otherwise its validity will be impugned and the original question becomes meaningless. For if any proffered standard of judgment may be invalidated by some other, none has any claim to rational allegiance. Conflicts, whenever or wherever they occur (and on the assumptions made they are likely to be ubiquitous), could never be resolved on any rational basis. The result is ethical scepticism and the inevitable consequence, an unrestricted license to individual inclination. Any semblance of duty can then be explained away only as compulsion by superior force, so that the only relevant motives for action become fear and self-interest. When everybody follows inclination indiscriminately, the result is anarchy; and it is well known that anarchy leads immediately to tyranny. Justice, in Thrasymachus' phrase, is then the will of the stronger, and the underpinning of morality is struck away. With it go both the motivation and the subject matter of ethics.

As the reference to Thrasymachus indicates, this state of affairs was not unknown to the ancients, although it arose for them from different scientific assumptions, and Plato and Aristotle countered it by appeal to standards established in the *polis*. Some modern philosophers, like Bradley and T. H. Green, with the help of Kant and Hegel, essayed to do the same in the early decades of this century by following the Greek tradition. But today this will not serve because the contemporary scientific outlook finds the Greek ideal, like all the rest, to be relative to a special type of culture so that it may well be (and indeed is) rejected by many outside of and untouched by Western European influence. The contemporary scene is therefore once again marked by conflict of voices, perplexity, and indecision.

Moreover, there is another outgrowth from the Newtonian presuppositions, still more ominously menacing today. It has already been noticed that the conception of the world as a machine, and the assumption that its "objectivity" is conditional upon its stark materiality and externality to mind, leaves nature outside the sphere of morality and sees it as neutral to value, except as providing means to human ends. The exploitation of nature is then not merely unexceptionable but actually applauded. Science, having discovered the laws of physics and chemistry, has enabled technology, the application of science for practical purposes, to exploit this knowledge to mechanize and industrialize our civilization. The recent consequences of this exploitation have become painfully obvious worldwide, in the expansion of deserts through overcultivation of land, the destruction of tropical forests essential to the world's oxygen supply, the depletion of the protective ozone layer in the upper atmosphere, the threatened exhaustion of energy resources, pollution of the air, the sea, rivers, and lakes, breaking the food chain vital to human sustenance, and other deleterious changes in the environment detrimental to the maintenance of life. Human interference with natural processes, both causing and attempting to counter these effects, promises even more disturbing developments involving genetic engineering, the release into the environment of new and hitherto unforeseen viruses, and the invention of new animal species.

With the help of science men have, in our day, acquired powers that were formerly attributable only to God. They can release the ultimate forces of nature in atomic fission and fusion. They can create new forms of life. They can (or may soon be able to) prolong human life by postponing the aging process. They can penetrate into outer space. They can alter the composition of the earth's envelope in ways that may have incalculable results for life on the planet. But what, as yet, they cannot do is decide how, or whether, these powers ought to be used. What science and technology can do now will have portentous consequences lasting into the unforeseeable future, in some cases, where radioactivity is concerned, for many thousands of years. So present acts will affect future generations as nothing that was possible in the past could affect present generations. Who may legitimately assume responsibility for such formidable acts? Although men may now have virtually divine powers, they have not the impeccable benevolence or omniscience of God.

The vital questions raised by these capabilities are no longer scientific. They are not merely how such results can be achieved, or how they can be counteracted, but whether such capacities as science and technology put at our disposal today should be exercised at all, or if at all under what controls. These, in the last resort, are moral questions. They are questions of priorities, which look to standards of value for their answers. Nor may these standards be merely local or provincial; they must be global because the exercise of such capacities can no longer be controlled by the powers of existing national authorities whose jurisdiction the consequences of that exercise transcend. Such control requires a worldwide authority, which, as yet, does not exist, as well as the ordering of priorities as to what is globally desirable, impossible without reference to some universal standard. But, because of the continuing influence of Newtonian presuppositions about the nature of the external world and of scientific knowledge, no such standard is recognized. We have been deprived by modern social anthropology and ethics of what we most desperately need.

iv

The demands of practical reason at the present time go far beyond the questions that formerly prompted ethical speculation. Hitherto philosophers have asked what was the best way for the individual to conduct his or her life. The conspectus of ethical inquiry was limited to the individual and the immediate community. It has never, until now, gone beyond the national state. Today the crucial questions concern the welfare of the entire human race and the prospects for future generations; and not only these but also the integrity of the whole planetary environment. These questions are unprecedented for ethical investigation, and its purview needs to be extended far beyond its former bounds. In fact a new approach is urgently demanded, and perhaps an altogether new type of ethics. On what foundation, one may ask, could it be built? There is, indeed, such a foundation, in spite of the present dismal appearances, because contemporary physics is no longer Newtonian. A new scientific revolution occurred at the beginning of the twentieth century, the implications of which have been largely neglected by philosophers and even by scientists in other disciplines than physics, although these implications for metaphysics and ethics are profound.

Contemporary physics has abandoned mechanism and even materialism. It no longer views the physical universe as a collection of material points in roughly external relations, moved by forces impinging on them from without. The current view of the physical world is of a single, indivisible, space-time whole, in which no separation can be made between space-time and energy, or between energy and matter; and in which the distinguishable elements are governed, both in their nature and activity, by the structure of the whole.

This conception is promulgated by all the major physicists of the age, from Max Planck and Albert Einstein to David Bohm and Fritjof Capra. They include Sir Arthur Eddington, Sir Edmund Whittaker, Louis de Broglie, Erwin Schrödinger, D. W. Sciama, Werner Heisenberg, Paul Davies, and Henry Stapp. The holistic outlook is derived from the requirements as much of relativity theory (special and general) as from the quantum theory, and it dominates astrophysics and cosmology as much as it does particle physics and atomic theory. In every province of physics, field and the energy system take precedence over particles and forces (where, as in relativity theory, the latter have not been eliminated in favor of geodesics), so that the structure of the whole regulates the character and the behavior of its parts.

From atomic theory holism invades crystalography and biochemistry, and thus the independent evidence in biology for organic integrity is reinforced, notwithstanding the persistence of reductionist prejudices among some biologists. Reductionism in biology today only leads back to the holism of particle physics, and the atomism of the Newtonian era has become altogether obsolete. Meanwhile, contemporary biology supports a similar change of outlook. The organic integrity of the living being has been recognized ever since Aristotle, and modern attempts to explain it away in mechanistic terms founder because reduction of metabolism to chemistry and chemistry to physics merely reintroduces into biology the holism of the physical system. But apart from this, contemporary biology provides copious evidence of the organismic character of regenerative processes, of the genetic code, of the genome, and of ontogeny. It has demonstrated the organic relationship of organism to environment, so that ecologists are at one in their insistence on the interdependence of species and individual organisms, of flora and fauna, in symbiotic communities, ranging from a droplet of water to a lake or an estuary, from the notch in a tree trunk to the whole tree and the entire forest, or from a coral reef to an ocean. In fact, they find that they cannot stop short of the planetary biosphere, the integrity of which is now seen to be vital to the survival of the human race. Further, the biosphere of the planet Earth is dependent on the constant supply of energy from the sun, which is part of the hierarchy of the physical universe in the galaxy and outer space beyond. So the holism of the biosphere is inseparable from the holism of the physical world.

This change in the presuppositions of science from a particulate to a unified world heralds a new metaphysic, a new logic, and a new ethic, derived from the nature of system. Thinking must be switched from the merely formal, based on what Hobbes called addition and subtraction of the implication of names, isolated atomic meanings, to the dialectical, deriving from the concrete self-differentiation of the whole. In ethics the search for a criterion must now look to the totality and not simply to the individual. We must take our cue from Plato and seek in the larger letters of the cosmos for clues to the virtues and functions of its members. The principle of organization governing the unified world will then be seen as universal, and its self-specification' in nature as issuing in its awareness of itself in human consciousness, as reason, which is objective for everybody, just because it is the principle universal to the whole. And it is in this principle that the source of social and moral order is to be sought. But this reason is no mere calculus serving ends set by capricious desires. It is the reason implicit in the cosmic totality, which asserts itself in the environment that determines the human condition, and is exercised in the conscious deliberation that adjusts human action in society to that environment. Such adjustment must be prompted by a new attitude to nature: no longer one of exploitation but rather one of tendance and cultivation. The maintenance of integrity now becomes paramount, not just of the person (important though that remains) but concomitantly of the biosphere. The tendance of the soul enjoined by Socrates is now linked with and conditional upon the tendance of the living world. And to this end we must look to political organizations which are not limited to national communities but encompass the entire globe.

Notes

1. Isaac Newton, *Mathematical Principles of Natural Philosophy*, trans. A. Motte, rev. F. Cajori (Berkeley: University of California Press, 1962), p. 544.
2. René Descartes, *Discourse on the Method*, trans. F. E. Sutcliffe (Harmondsworth: Penguin, 1968), p. 64.
3. Thomas Hobbes, *Leviathan*, ed. C. B. Macpherson (Harmondsworth: Penguin, 1968), chap. 46, p. 689.
4. Ibid., introduction, p. 81.
5. Ibid., chap. 1, p. 86.
6. G. E. Moore, *Principia Ethica* (Cambridge: Cambridge University Press, 1903).
7. Bertrand Russell, "Reply to Criticisms," in *The Philosophy of Bertrand Russell*, ed. P. A. Schilpp (Evanston, Ill.: Open Court, 1946), p. 723.
8. Kurt Baier, *The Moral Point of View* (Ithaca: Cornell University Press, 1958).
9. Alasdair MacIntyre, *After Virtue* (Notre Dame: Notre Dame University Press, 1981).
10. F. H. Bradley, *Ethical Studies*, 2d ed. (Oxford: Clarendon Press, 1927).

Chapter 13

Modern Ethical Theory and Newtonian Science: Comments on Errol Harris

Philip T. Grier

i

Errol Harris reminds us that in the history of ethical theory we are confronted with a significant distinction between what we generally recognize to be "modern" ethical theory on one hand and "premodern" on the other. As he suggests, the triumph of Newtonian science, or more precisely the triumph of mechanist cosmology which it sponsored, appears to be the single most important dividing line that separates the two. In our retrospective picture of these developments it appears that the emergence of Newtonian mechanics, with its attendant philosophical baggage, disrupted an established tradition of ethical thought and placed certain recognizable constraints on all subsequent ethical theories, by which signs we know them to be "modern."

The most distinctive signs of this "modernity" in ethical theory are, as Harris implies, subjectivism,[1] voluntarism,[2] and relativism.[3] (I am deliberately omitting naturalism, another obvious candidate for this list, on the grounds that within the modern tradition "a naturalist ethics" seems to me more like another name for the problem rather than a possible solution to it). As he points out, these three characteristics apply to an impressively large part of ethical theory since the seventeenth and eighteenth centuries. They have been more or less willingly embraced by Hobbes, by Hume and the moral sense theorists, as well as by emotivists, prescriptivists, and other noncognitivists and existentialists of the twentieth century. If we subtract voluntarism from the list, then the remaining two characteristics arguably apply to utilitarianism in its various forms. (Ideal Observer theories are a slightly more complex case.)

Harris' account presents all three of these characteristics of modern ethics as rather direct consequences of Newtonian science. A subjectivist account of value appeared inevitable, given the new philosophy of nature involved in Newtonian mechanics. Subjectivist accounts of value seemed to entail a demotion of the role of reason in the making of moral judgments, clearing the way for ethical voluntarism. These two developments in turn, having thoroughly excluded moral or other valuations from the sphere of objective science, cleared the way for the emergence of "value-free" descriptive sciences of society and human nature. They in turn produced descriptive theories of cultural relativism, which could be cited as "justifications" for the truth of ethical relativism. Thus the general situation of modern ethical theory was set.

There has, of course, also been a countertrend in modern ethical thought which can be identified by its struggle to avoid these outcomes of voluntarism, subjectivism, and relativism, usually by means of an objectivist account either of moral duty or of the good. As Harris indicates, this countertrend would include Kantian ethics and the various nonnaturalist, intuitivist theories, such as Moore's account of the good and the deontologisms of Ross and Prichard. To this list we could add others such as Scheler and Nicolai Hartmann. In Harris' view, despite laudable intentions, all these ethical theorists fail in one way or another to establish the objectivity and universality of moral judgments which they profess. And he traces these failures for the most part to a continued deference to what are widely thought to be inevitable consequences of Newtonian mechanist cosmology.

It is clear that over the last several years a significant dissatisfaction with both sides of this entire modernist tradition of ethical theory has set in, and we are already well launched into the project of what has been nicely described as "the recovery of premodern ethics." Alasdair MacIntyre's *After Virtue* represents one of the most visible strands of this effort, in both its critique of the modern ethical tradition and its explicit attempt to reestablish the plausibility of an Aristotelian community-based ethic of individual character.

A parallel effort to restore a character-based ethic has also emerged in the context of political philosophy. This takes the form of a critique of the liberal (especially the Kantian) conception of the self as having a moral identity independent of all group memberships and communities, and conceived as prior to any of its particular ends or values (save perhaps the one value of being free to choose or to abandon any other end or value as I see fit). Among the critics of this Kantian conception of the self are to be found Michael Walzer and Michael Sandel, in addition to MacIntyre.[4]

The resurgence of interest in Hegel's ethics, with its significantly Aristotelian emphasis on the centrality of ethical community, could also be cited as a contribution to this same movement.[5]

A second direction of effort in this recovery of "premodern" ethics has been visible for a number of years in the attempted recovery and defense of natural law ethics. I have in mind here especially the work of Germain Grisez, and also John Finnis, whose *Natural Law and Natural Rights* was significantly influenced by Grisez.[6]

In my response to Professor Harris' essay I will connect his views primarily with this second direction of thought, the recovery of the natural law tradition, for three reasons. First, to the extent that we think of mechanist cosmology as disrupting an established tradition of ethical thought in the seventeenth century, that tradition was predominantly one of natural law ethics.

Second, Harris' critique of the "modernist" tradition in ethical theory treats voluntarism, subjectivism, and relativism as unacceptable deficiencies. By implication, then, an adequate ethical theory in Harris' view would have to be rationalist[7] as opposed to voluntarist, objectivist[8] as opposed to subjectivist, and universalist[9] as opposed to relativist. These three characteristics (rationalism, objectivism, and universalism) apply most obviously to the natural law tradition and would seem to imply that Harris' own position belongs at least generically to that tradition.

Third, at the end of his essay, Harris does indeed seem to be sketching what could be described as a new natural law ethic for the future which bears some characteristic marks of the traditional theory (though in the context of a new philosophy of nature, a new metaphysic, and a significantly altered account of theoretical reason).

ii

In general I do not propose to disagree very much with Professor Harris' view. It seems to me that the account he sketches of the consequences of Newtonian science for ethical theory and also for conceptions of human nature and society up through the Enlightenment is quite correct. In the period following the Enlightenment, though, I am tempted to add some complications, especially in the area of conceptions of human nature, taking into account some of the reactions against the Enlightenment. I will suggest that from the end of the eighteenth century on, some of the special characteristics of modern ethical thought which we have attributed to the influence of Newtonian science begin to acquire alternative sources of support and a momentum of their own. Some of these developments also involve a somewhat changed appearance for discourse concerning the ethical. To do this I will broaden the subject slightly from ethical theory to practical reason and its relation to theoretical reason. I want to focus on changing images of reason, both theoretical and practical, in the modern period, trying to distinguish between changes that appear to be direct consequences of the success of Newtonian mechanics and those that appear to have other sources.[10]

Turning to the first of the three characteristic features of modern ethics, subjectivism, it seems to me that Harris is surely correct in claiming that the most direct impact of Newtonian science on ethical objectivism was due to radical changes in the philosophy of nature brought about by the introduction of a new conception of matter. The implications of that development for ethical theory are fully drawn out. I would only add that his conclusions are very well supported by Ivor Leclerc's work *The Philosophy of Nature*.[11] Leclerc argues that this new philosophy of nature should be seen as a distinctively seventeenth-century version of Neoplatonism, evolving from Renaissance Neoplatonism, in the work of Sebastian Basso, Galileo, Descartes, and Newton. Its most distinctive feature was the doctrine of matter conceived as a substance in its own right, a radical modification of the Aristotelian doctrine of substance in which form and matter were correlative. Separated from form, matter as substance was conceived to be fully in being, incapable of becoming, therefore without capacity for internal development and equally incapable of having its essence as matter changed from without. Given this conception of matter, change or motion in nature could be conceived only in terms of change of place (locomotion) of changeless matter.[12] Such a conception of matter as substance necessarily implied a dualism of mind and nature. Moreover, since matter, always fully in being, could no longer be conceived as developing from potentiality toward the actuality of perfected form, there was apparently no longer any basis for grounding value in nature.

To Professor Harris' account of the impact of Newtonian science on ethical objectivism I would add the following considerations: If values and moral principles were no longer conceived as grounded in nature in any sense, then the claim of practical reason to provide objective knowledge of them might seem beside the point. Nevertheless, there were vigorous attempts in the seventeenth century to sustain this claim that practical reason could supply objective moral knowledge even in the absence of any teleological conception of nature. The reformulations of natural law theory by Grotius and Pufendorf were intended to accomplish this. And the Cambridge Platonists defended similar positions in their reactions to Hobbes's attack on the objectivity of ethical judgments. In all of these cases human nature was more or less identified with the possession of reason, and practical reason was held to be capable of delivering objective knowledge of universal, self-evident moral principles.[13]

It seems to me that this attitude toward practical reason came under inexorable pressure as a consequence of changes in attitude toward the power of *theoretical* reason brought about by the success of Newtonian science. The changes in question did not come about immediately, but over the course of the seventeenth and eighteenth centuries a dramatic change in attitudes about the power of theoretical reason did take place.

In a simplified summary, one could have held at the outset of the seventeenth century that theoretical (that is, speculative) reason was capable of intuiting universal and necessary (mathematical) propositions governing the motions of things in nature, and thus natural philosophy (or physics) could be thought to produce demonstrative knowledge based on self-evidently true propositions. (I do not mean to claim that Newton's conception of scientific reasoning could be fitted into this mold, but Descartes, for example, was at least ambivalently committed to such a view.)[14]

Analogously, practical reason could be held capable of intuiting universal moral principles grounded in the nature of things, reasoning out their implications in particular circumstances, and arriving at objective conclusions of conscience. Both science and morality could thus be presented as complementary sources of objective knowledge derived from different applications of the one faculty of pure reason. Under these circumstances, the claim of practical reason to supply objective knowledge could derive strong support from the comparison with theoretical reason.

By the end of the eighteenth century theoretical (that is, speculative) reason was generally held to be involved only indirectly at most in the objective knowledge of nature; that knowledge was thought to be an affair of natural reason, or the understanding; it was not thought to involve necessarily true principles; and nature conceived as thing-in-itself was thought to be not knowable at all. This drastically revised conception of theoretical reason now offered scant support for a view of practical reason as capable of supplying objective moral knowledge. (Kant supplied the most heroic effort along these lines but of course could not claim to know the truth of his postulates of practical reason.)

Thus it was not just the Newtonian philosophy of nature that had direct ethical implications but also the impact of Newtonian science on our conception of theoretical reason and its role in obtaining objective knowledge. The latter development was equally important, I would suggest, in depriving natural law ethics of plausibility, especially the seventeenth-century versions of natural law theory mentioned earlier.

iii

A second qualification or supplement to Harris' account that I would suggest concerns voluntarism. It seems to me that voluntarism is a development that has somethng of a life of its own in the modern period, to some extent emerging independently of the impact of Newtonian science.

Voluntarism—the view that goals and ends, as products of human will, are not subject to rational assessment, that the will is superior to intellect or reason—was undoubtedly encouraged by the dominance of mechanist cosmology. That is, voluntarism is closely linked to subjectivism, and the latter was strongly promoted by Newtonian cosmology. Nevertheless, voluntarism constitutes a distinctive attitude toward the role of reason in human affairs, and the implicit question it poses—In what does the greatest source of human dignity lie: in the exercise of reason or the exercise of will?—is of great consequence for philosophical anthropology. It is clear that some thinkers find in voluntarism a distinct source of human dignity, and because of this voluntarism seems to have something of a life of its own, independent of the influence of Newtonian science.

The emergence of this attitude, which eventually leads to a distinctly post-Enlightenment conception of human nature, can be briefly traced as follows.

In the medieval period the human capacity to love God and also to know God's word and works was seen as the chief source of human dignity. Beginning perhaps with Nicholas of Cusa and his conception of the world as *explicatio Dei* there was a distinct emphasis on knowledge of nature as a path toward the knowledge of God, even if the ultimate nature of God was beyond human ken. If Richard Popkin is correct in his recent work identifying seventeenth-century millenarianism as an important motive behind much work in seventeenth-century natural science, including that of Newton, then the picture that emerges of attitudes toward human reason, its powers and limits, may be less changed during the seventeenth century than is often supposed.[15] It may not have been until later in the eighteenth century that knowledge of God's creation was generally abandoned as a tacit motive of the scientific investigation of nature and this understanding of the ultimate ground of human dignity gradually modified.

By the end of the eighteenth century, however, with widespread acceptance of the claim that human reason is intrinsically incapable of knowing nature as it really is (the thing-in-itself), Kant offers what amounts to a different ground for the ultimate in human dignity, not a form of knowledge at all but rather the capacity to act in accordance with the moral law. The greatest source of human dignity is thus grounded by Kant not in theoretical but in practical reason, in morality rather than science.

Kant's treatment of human moral capacities in such elevated terms was of course greatly and immediately influential, especially in German literature and philosophy. But within surprisingly few years yet another account was provided of the ultimate source of human dignity, the "highest exercise of human reason," namely *art*. In the famous short document called "The Earliest System-Program of German Idealism" (Berne, 1796), the authorship of which was at various times attributed to Schelling, to Hölderlin, and to Hegel, it is declared that in writing a philosophical history of humanity and all of its works, the culminating Idea that unites all the rest must be the Idea of *beauty* in its higher, Platonic sense. "The highest act of Reason, the one through which it encompasses all Ideas, is an aesthetic act, and . . . *truth and goodness only become sisters in beauty*—the philosopher must possess just as much aesthetic power as the poet . . . One cannot be creative [*geistreich*] in any way, even about history one cannot argue creatively—without aesthetic sense." "Poetry gains thereby the higher dignity, she becomes at the end once more, what she was in the beginning—the *teacher of Mankind*."[16] This notion of artistic creativity as the highest act of reason, the highest ground of human dignity, was of course celebrated in various forms of romanticism.

The notion of artistic creativity as the highest ground of human dignity is subtly but significantly transformed in Marx. The focus of creativity is no longer on the act of individual artistic genius but rather on the possibility of humanity as a whole (one declared to be divine by Feuerbach) creating a new, authentic form of human nature through the creation of a new social world. Creativity is still the ground of human dignity, but it is now the God-rivaling act of creation itself.

For Nietzsche, too, creativity is the ultimate ground of dignity, but of course it is the possibility of the creation of new values that is celebrated, and not "human" values but the values of the overman.

It is not entirely clear to me whether this particular path of cultural evolution comes to an end in the existentialisms of the mid-twentieth century, or whether we are still witnessing various forms of the same development in Derrida and Foucault. It is fairly clear, however, that from Marx onward the notion of creativity as the ultimate ground of human dignity is no longer explicitly connected or connectable to the theme of reason except in this transformed sense. This too, then, could be viewed as a demotion of practical reason no less significant than that brought about by the triumph of Newtonian science, but not a direct consequence of it.

More important, is seems to me that the dominant philosophical anthropology of the twentieth century, insofar as there can be said to be one, is rooted in this nineteenth-century doctrine of expressivism, with its celebration of human will, of creative action. From within this tradition the problem of the nature of practical reason may look rather different than from without (as different, say, as Nietzsche is from Bishop Butler).

iv

Turning to the third dominant characteristic of modern ethical theory, relativism, again it seems to me that there are some additional factors beyond the effects of Newtonian science which need to be brought into the picture, again having to do with an alternative paradigm of reason as well as with a significant development in philosophical anthropology.

There is of course another distinctive paradigm of reason which emerges in the late eighteenth century, inspired partly by Rousseau, but most influentially expressed by Hamann and Herder. That alternative paradigm of reason holds that it neither originates in nor is modeled after abstract mathematical patterns or other universal formalizable structures of thought. Rather it originates in, and cannot finally be separated from, natural language.

Natural language, according to Hamann, serves as the "womb" of reason and of all artificial constructions designed to assist in reasoning. In natural language the operations of reason and experience cannot be separated out, and the thinking that occurs in natural language takes precedence over all specialized or technical applications of language such as philosophy and science.[17] Similarly, Herder argues that the operations of reason and the use of natural language are inseparable.[18]

It was this paradigm of reason that lay behind the new "expressivist" anthropology of German idealism and romanticism, which has been so well described by Charles Taylor.[19] As he points out, Herder, in describing man as a rational animal, was significantly transforming the ancient notion. Human activity and human life were seen as expressions of meanings which were to some extent brought into being as acts of self-definition by individuals or peoples who, in their language and art, gave objective form, actuality, to what was, prior to being expressed, only potentially their being. "Man comes to know himself by expressing and hence clarifying what he is and recognizing himself in this expression. The specific property of human life is to culminate in self-awareness through expression."[20] Man is for Herder a rational animal, but "rationality is not a principle of conformity with cosmic order. Rather it is self-clarity, *Besonnenheit*. Achieving this, we become what we have it in us to be, we express our full selves, and hence are free."[21]

There is obviously an important implication for practical reason lurking here. If natural language is taken as the paradigm of reason, instead of some explicitly universal structure of thought such as mathematics, does this not entail that the scope of reason is now culturally bound and hence incapable of delivering universal truth, but only truth relative to a culture? (From a Hegelian perspective the answer is of course finally no, but as his account of *Sittlichkeit* makes clear, the connection between particular national cultures in history, and universal ethical truths, is a very complicated affair.)

This alternative paradigm of reason, with its implication that truth is relative to particular cultures, particular languages, is seldom put in such simple fashion in the twentieth century, but allied with historicism, its influences can still be detected. There is an alternative philosophical tradition involving aspects of the late Heidegger, and especially the hermeneutics of Gadamer in which much of the problematic of Hamann, Herder, and Humbolt concerning the role of language in human existence can still be discerned.[22] This tradition has been all but mute on the subject of formal ethical theory, but it lays claim to much of the same territory in a very different guise. From within this tradition, with its distinctive attitudes toward the connection between language and reason, the situation of practical reason appears still more convoluted and remote from the problematic of contemporary science.

v

In contemporary terms the problem finally becomes whether practical reason has the resources to establish objectivity of some sort for moral judgments, or alternatively, whether and in what sense this concern for objectivity is thought to have any significance. Dealing only with the first issue at present, four different responses to this problem are of interest here.

In his recent book *Whose Justice? Which Rationality?* Alasdair MacIntyre has arged that no doctrine of practical rationality can be expected to satisfy contemporary science-derived standards of rational justification.[23] He looks skeptically at the Enlightenment aspiration "to provide for debate in the public realm standards and methods of rational justification by which alternative courses of action in every sphere of life could be adjudged just or unjust, rational or irrational, enlightened or unenlightened,"[24] denying that any such standard of universal rationality is available to us. Instead, he argues that all conceptions of justice and practical rationality must be seen to operate only within and relative to some particular "socially embodied" tradition, and derive their validity only from such sources. He examines four such traditions of practical rationality—those associated with Aristotle, Augustine, Aquinas, and Hume—then goes on to explore the question of the extent to which the existence of a plurality of distinct traditions of practical rationality can serve as the basis for any more general notion of practical rationality, in terms of the possibilities for dialogue among such traditions.

Germain Grisez and John Finnis have both attempted to respond to the potential conflict between modern scientific notions of rationality and their doctrines of practical rationality by means of a different strategy. They attempt to distinguish very sharply between what they refer to as the metaphysical context on one hand and the actual principles of "practical reasonableness" on the other.[25] In this way they appear to hope that when particular assumptions belonging to the metaphysical context appear to collide with contemporary scientific theory, the damage can be seen not to extend to their doctrines of practical reason. Their strategy is thus to insist on a kind of autonomy of practical reason, grounded in self-evidence and unaffected by the contemporary adventures of theoretical reason.

That strategy has been subjected to criticism in another book, *A Critique of the New Natural Law Theory*, by Russell Hittinger.[26] Hittinger rejects this attempted separation of practical reason from theoretical reason, arguing that unless one brings both philosophical anthropology and metaphysics into the account, one cannot show how or why "nature" should be normative for human activity. He concludes that if natural law theory is to be recovered, the divorce between theoretical reason and practical reason must be transcended.[27]

That would seem to be a very tall order, but Professor Harris' sketch of a possible new ethic in his concluding remarks can be seen as an answer to that very demand. He argues that the possibility of overcoming the divorce between theoretical and practical reason is at hand, a possibility arising from a specific interpretation of the philosophical implications of much twentieth-century science. That interpretation involves a new philosophy of nature, a new anthropology, and a dialectical account of the structure of theoretical reason.[28] The philosophy of nature involves a necessary principle of development, or rather a dialectically self-developing hierarchy of such principles in terms of which one can give an account of value once again based in nature. Practical reason would again be characterized by rationalism, objectivism, and universalism. In these respects it seems to me that the new ethic toward which Harris points would be a new version of natural law ethics.

As for the likelihood of such a development, I must confess to certain doubts. First of all it seems to me that, independently of the emergence of Newtonian science, one can argue that in much of modern intellectual history we can see a preference for voluntarism being exercised, perhaps as a kind of ultimate conceptual barrier against all authorities who would claim, on the basis of superior knowledge, to dictate to us concerning the proper ends of human life. Some even claim to see greater human dignity in a situation where individual humans must be responsible for creating their own values (whatever that means).

One could also point to a general fragmentation of contemporary culture, some of which I have just been sketching. Any wide acceptance of a new concept of practical reason closely unified with theoretical reason seems to me to presuppose first some sort of general reunification in the cultural sphere, a New Age. But perhaps we are on the threshold of one.

Notes

1. By "subjectivism" I mean the claim that ethical assertions always depend in a crucial way on the attitudes or feelings of the person making the assertion.

2. By "voluntarism" I mean in general the view that the human will is superior to intellect or reason, and more specifically the view that human ends and values are products of the human will and as such are not subject to rational assessment, are neither rational nor irrational as such.

3. By "relativism" I mean the claim, sometimes referred to as "normative relativism," namely, that an individual is morally obligated to perform, or refrain from performing, an act if and only if in the particular society of which that individual is a member the act in question is morally required, or condemned. The view in question would normally presuppose the truth of cultural (descriptive) relativism.

4. See Michael Walzer, *Sphere of Justice* (Basic Books, 1983), and Michael Sandel, *Liberalism and the Limits of Justice* (Cambridge University Press, 1982). See also Sandel's very useful anthology *Liberalism and Its Critics*, ed. Michael J. Sandel (New York University Press, 1984).

5. Other attempts to recover elements of premodern ethical thought have been made by G. E. M. Anscombe, Alan Donagan, Stanley Hauerwas, and others.

6. John Finnis, *Natural Law and Natural Rights* (Clarendon Press, 1980). Grisez's writings on the subject are numerous; they include such items as Grisez and Shaw,

Beyond the New Morality (University of Notre Dame Press, 1980); Grisez, "The First Principle of Practical Reason: A Commentary on the *Summa Theologica*, 1–2, Question 94, Article 2," *Natural Law Forum* 10, reprinted in A. Kenny, ed., *Aquinas: A Collection of Critical Essays* (London, 1970); G. Grisez, "Presidential Address: Practical Reasoning and Christian Faith," *Proceedings of the Catholic Philosophical Association* 58 (1984), 2–14.

7. I mean the view that moral ends and values are subject to rational assessment and validation.

8. The claim that the truth of ethical assertions does not depend uniquely on the attitudes or feeelings of the person making the assertion but on features of the situation and principles which are assessable by any rational observer.

9. The claim that the validity of general moral principles and values is not relative to specific cultural groupings but applies to human beings as such.

10. This is a somewhat delicate undertaking, because we tend to use conceptions of reason as summative notions, to distinguish one philosophical age from another, and only retrospectively. The philosophical battles that give rise to changing conceptions of the powers and limits of human reason are usually fought on narrower and firmer ground, in theology or science or metaphysics or epistemology, or philosophical anthropology. Nevertheless, we are engaged in a retrospective inquiry here, and I think this approach may introduce some appropriate complications in the picture presented by Professor Harris.

11. Ivor Leclerc, *The Philosophy of Nature* (Catholic University of American Press, 1986).

12. Ibid., p. 5.

13. For example, Cudworth and More maintained that moral principles are self-evident truths as certain and universal as the truths of mathematics and, like the latter, discoverable by the exercise of pure reason.

14. For example, "I only consider [in physics] the divisions, shapes and movements [of quantity, regarded geometrically]; and I do not want to receive as true anything but what can be deduced from these with as much evidence as will allow it to stand as a mathematical demonstration." Descartes, *Principles* II, 64.

15. See Richard Popkin, "The Religious Background of Seventeenth-Century Philosophy," *Journal of the History of Philosophy* 25, no. 1 (January 1987), 35–50.

16. Quoted from the translation by H. S. Harris, in his *Hegel's Development: Toward the Sunlight, 1770–1801*, appendix, p. 511.

17. See, for example, Hamann's *Aesthetica in Nuce*, in H. B. Nisbet, trans. and ed., *German Aesthetic and Literary Criticism: Winckelmann, Lessing, Hamann, Herder, Schiller, Goethe* (Cambridge University Press, 1985).

18. See James C. O'Flaherty, *Johann Georg Hamann* (Twayne, 1979), esp. chap. 7. See also W. M. Alexander, "Hamann, Johann Georg," in Edwards, ed., *Encyclopedia of Philosophy*, vol. 3.

19. In Charles Taylor, *Hegel* (Cambridge University Press, 1975), chap. 1.

20. Ibid., p. 17.

21. Ibid., p. 22.

22. I have in mind especially the third part of Gadamer's *Truth and Method*, "The ontological shift of hermeneutics guided by language."

23. Alasdair MacIntyre, *Whose Justice? Which Rationality?* (University of Notre Dame Press, 1988), esp. chaps. 1, 19, and 20.

24. Ibid., p. 6.

25. For example, Finnis: "Thus it is simply not true that 'any form of natural-law theory of morals entails the belief that propositions about man's duties and obligations can be inferred from propositions about his nature.' Nor is it true that for

Aquinas 'good and evil are concepts analysed and fixed in metaphysics before they are applied in morals.' On the contrary, Aquinas asserts as plainly as possible that the first principles of natural law, which specify the basic forms of good and evil and which can be adequately grasped by anyone of the age of reason (and not just by metaphysicians), are *per se nota* (self-evident) and indemonstrable. They are not inferred from metaphysical propositions about human nature, or about the nature of good and evil, or about 'the function of a human being,' nor are they inferred from a teleological conception of nature or any other conception of nature. They are not inferred or derived from anything. They are underived (though not innate). Principles of right and wrong, too, are derived from these first, premoral principles of practical reasonableness, and not from any facts, whether metaphysical or otherwise. When discerning what is good, to be pursued (*prosequendum*), intelligence is operating in a different way, yielding a different logic, from when it is discerning what is the case (historically, scientifically, or metaphysically); but there is no good reason for asserting that the latter operations of intelligence are more rational than the former." *Natural Law and Natural Right*, pp. 33–34.

26. Russell Hittinger, *A Critique of the New Natural Law Theory* (University of Notre Dame Press, 1987).

27. See especially his conclusions, pp. 193–196.

28. Professor Harris' work on the philosophical implications of contemporary science extends over many publications and many years. Three of the most important works presenting his thought on these problems are *The Foundations of Metaphysics in Science* (George Allen and Unwin, 1965; reprint, University Press of America, 1983); *Hypothesis and Perception: The Roots of Scientific Method*, Muirhead Library of Philosophy, H. D. Lewis, ed. (Humanities Press, 1970); and *Formal, Transcendental, and Dialectical Thinking* (SUNY Press, 1987).

Contributors

Peter Achinstein
Johns Hopkins University

Richard T. W. Arthur
Middlebury College

Phillip Bricker
University of Massachusetts
at Amherst

John Carriero
Harvard University

Robert DiSalle
University of Western Ontario

Michael Friedman
University of Illinois at Chicago

Philip Grier
Dickenson College

Errol Harris, Emeritus
Northwestern University

R. I. G. Hughes
University of South Carolina

J. E. McGuire
University of Pittsburgh

Lawrence Sklar
University of Michigan

Howard Stein
University of Chicago

Index

absolute motion. *See* motion, absolute
absolute place, 58–59, 66–67, 84
absolute space or time. *See* space, absolute; time, absolute
absolutism. *See* realism, about space and time
acceleration, 23–24, 203, 205, 208–209. *See also* inertia; motion
absolute, 59, 70–71, 73, 78–80, 84, 86–87
and gravity, 189–195, 203, 208
and inertial effects, 64, 66, 78
as primitive property, 86–87, 89
as a property of material objects, 79
relative, 66, 70
accidents, proper, 112, 122–124, 127–128. *See also* space and time; substance
Achinstein, Peter, 8–9, 175–183
action at a distance, 53, 55, 56n3. *See also* impact; gravity, as immediate
active principles, 10, 38–39, 53, 208. *See also* gravity; passive principles
actuality, relation to space and time, 93, 101, 103, 106
aetherial medium, 151, 186, 188–190, 193–194, 206
affections of being *qua* being, space and time as, 102–106, 109–111, 115–120, 124
Aquinas, St. Thomas, 112, 121, 124–127, 236
Aristotle, 2, 3, 49, 236
on actuality, 94, 101–102
on autonomy of the moral self, 219
ethics of, 228
on first principles, 33–34, 38, 51, 55
metaphysics of, 7, 109, 126, 219, 223
on morality, 221
on substance as form and matter, 230
on unity and being, 126
art, vs. reason as source of human dignity, 233–234
Arthur, Richard, 5
attraction and repulsion. *See* Kant, balancing argument of; gravity
Augustine, Saint, 93, 105, 236
Ayers, Michael, 33, 36

Baier, Kurt, 216–217
balancing argument. *See* Kant, balancing argument of, for essentiality of gravity to matter; gravity
Balmer's law, 54
Barbour, Julian, 15n13
Bayes's theorem, 162
being, as substance and accident, 109, 126. *See also* affections of being *qua* being, space and time as; existence
Bentley, Richard, 11, 13, 199
Berkeley, George, 28–29, 56, 185
Boethius, 93
Boyle, Robert, 9, 28, 151
Bradley, F. H., 220
Bricker, Phillip, 6
bucket, rotating, from Newton's Scholium to the Definitions, 58, 69, 71, 74, 79, 85–87
Burtt, E. A., 2, 5, 55–56

Capra, Fritjof, 12–14
Carriero, John, 7, 91, 103–106
causal inference, to existence of space and time, 60, 81–82, 87, 135–169, 175–183
causal explanation. *See* inertial forces, causal explanation for

causation, 135–169, 175–183, 218
  efficient, 7, 105–106, 111–115
  emanative. *See* emanative causation
  role of absolute quantities in, 60, 63–64, 70–75, 79, 81–88
  of space and time, 79, 81–88, 109, 111–115 (*see also* God, as cause of space and time)
centrifugal force, 20, 22–26, 50, 83. *See also* motion, circular
centripetal force, 139, 142–143. *See also* motion, circular
certainty, 33–34
  degrees of in experimental philosophy, 135–169
Clarke, Samuel, 7, 70, 103–104, 117–119, 122–124, 127–128
clocks, 58, 65–68, 70–71, 73–74, 78–88
  and rods, 61–65
Cohen, I. B., 95, 171n28
cohesion, 37, 53, 187–188
color, 27–29, 30, 39, 140–141, 145–146, 157–158, 175–176, 179–181
conservation of momentum, 20–26, 38, 50, 191, 193–194, 199, 205–209
Copernicus, 190–191, 198, 211, 219
corpuscularian philosophy, 19, 26, 29, 32–33, 36, 38, 49, 51, 54. *See also* light, particle theory of; mechanism, after Newton
corpuscular theory of light. *See* light, particle theory of
Cotes, Roger, 136, 194, 199
cultural relativism, as justification for ethical relativism. *See* ethical relativism

deductions, of propositions from phenomena, 9, 136, 139–169, 175–183. *See also* hypothesis; induction
Deism, 13
Descartes, Rene, 18–21, 26, 34, 56, 144, 150
  conception of God, 95–96
  on "first philosophy", 3, 34
  metaphysics of, 54, 198, 213, 214, 216, 218
  on motion, 7–8
  physics of, 18–21, 40n8, 49–52, 183
  Principia Philosophiae, 3, 18
  on reason, 231
design, argument from, 4, 13, 98

Des Maizeaux, Pierre, 93, 96, 101, 117–118, 122–124
diffraction, 157–158, 166
DiSalle, Robert, 10

Einstein, Albert, 77, 85, 209. *See also* relativity
emanative causation, of space and time by God, 7, 104–106, 111–128. *See also* causation, efficient
emotivism, 216, 218–219, 227
essence. *See* existence, vs. essence
eternity, 93–94, 100–106, 114, 119
ethical relativism, 215–216, 218, 219–220, 227–229, 234–237
ethics, Newtonian science and, 211–225, 227–237. *See also* morality; natural law
  deontology in, 214, 217–219, 228
  empiricist, 216
  ethical scepticism, 219–221
  new scope of, 222–225
  utilitarianism in, 214, 216–219
existence, 100–106
  as entailing space and time, 93–94, 116–128 (*see also* actuality)
  vs. essence, 100, 106, 109–113, 117–128
  not a substance or a quality, 101–102, 110, 118
experimental philosophy, Newton's, 135, 144–169, 175–183. *See also* hypothesis
expressivism, 233–235
extension, 19, 34–35, 100, 103, 109, 140, 144–145, 213

fact/value distinction, 213–214, 216, 219
Finnis, John, 236
first principles, 34, 38, 51–53, 55
free will, 211, 217
Friedman, Michael, 9–10, 203–209

Galileo, 3–7, 18, 50, 219
  laws of motion of, 20–24
  space-time, 203–205
Gassendi, Pierre, 18, 50, 52
General Scholium. *See* Newton, General Scholium
God
  actuality of, vs. essence of, 91, 102
  as cause of motion, 3, 32
  as cause of space and time, 7, 91–106, 111, 115, 122–124, 127–128

as creator, 36, 51, 53, 92, 97, 211–212
emanative effects of. *See* emanative
    causation
eternity of, 97–106, 119, 123, 127
existence of, arguments for, 3–4, 13, 92,
    98
existence and essence of, 100–106 (*see
    also* existence, vs. essence)
as identical with time and space, 104–
    106, 127–128
immanence of in the world, 13
infinity of, 94–106, 119, 123
Newton's anthropomorphic account of,
    94, 100
omnipotence of, 97–98, 201
omnipresence of, 13, 95–106, 119, 125,
    127, 201
omniscience of, 97–98, 201
omnitemporality of, 106
as (infinite) perfection, 95, 97–100, 106
search for knowledge of, as honor, 91–
    92, and dignity, 232
sempiternity of, 92–96, 98
as not timeless, 92–93
Goodman, Nelson, 71, 162, 176
gravity, 21–26, 32–33, 35–39, 46*n*88, 60,
    136, 140–143, 148, 151, 185–202, 203–
    209. *See also* active principles; laws of
    motion
as essential to matter, 185–202, 206–209
    (*see also* Kant, balancing argument of)
as immediate, 186, 192, 194–195, 198,
    201–202 (*see also* Kant, balancing argu-
    ment of)
Huygens' theory of, 21–26
as proportionate to mass, 188–200
as universal, 141, 183, 185–186, 189–
    202, 207
Grier, Philip, 10–11
Grisez, Germain, 236

Harris, Errol, 10–11, 227–237
Hegel, G. W. F., 228, 235
Hittinger, Russell, 236
Hobbes, Thomas, 150, 212–214, 217, 227
holism, patterns of in contemporary sci-
    ence, 11, 223–225
Hooke, Robert, 26, 30, 52, 151, 155
Hume, David, 12, 30, 56, 176, 214–215,
    227, 236

Huygens, Christian, 4, 17–26, 37–38, 49–
    56
and first philosophy, 34
theory of impact, 21–26
Treatise on Light, 26
wave theory of light, 54, 155
hypothesis, place of in experimental phi-
    losophy, 3, 8–10, 38–39, 52–55, 135–
    169, 175–183

impact, 19–26, 49, 52, 205. *See also* laws
    of motion, Newton's
impenetrability, as a quality of sub-
    stance, 35, 36–37, 51, 54, 136, 139–140,
    144–145, 187, 197
incitation, Huygens' theory of, 25, 37–38
induction, Newton's view of, 136, 140–
    146, 161–165, 169, 169*n*10, 175–183
inertia, 7, 21, 34–35, 37–38, 49–50, 53,
    58, 64, 66–67, 71, 73, 81, 164, 166, 191,
    200, 205–209
inertial forces, 49, 64, 66–67, 72–74, 81,
    205–207
causal explanation for, 78, 84–88
inference. *See* causal inference; deduc-
    tions; hypothesis; induction
infinity, 91, 99–103, 113. *See also* space,
    absolute
inverse-square forces, 37, 60, 142–143,
    189–191, 193. *See also* gravity

Jupiter, 137, 139
and Saturn, 141, 189, 193–195

Kant, Immanuel, 9–10, 12, 185–209
balancing argument of, for essentiality
    of gravity to matter, 186–188, 196–197
    (*see also* gravity)
Critique of Pure Reason, 10, 32, 185
empirical and noumenal causation of,
    218
ethics of, 228, 231, 233
*Metaphysical Foundations of Natural Sci-
    ence*, 9–10, 185–202, 204–205
Religion within the Limits of Reason Alone,
    201
Kepler, Johannes, 3, 37, 138, 142, 192
kinematics, 7, 181
Koyré, Alexander, 69

Lambert, J. H., 195
laws of motion, Newton's, 20–26, 37–38, 51, 60–61, 68, 111, 136, 185, 191, 203, 205, 207–209, 212–213. *See also* motion
Leclerc, Ivor, 230
Leibniz, G. W., 6–8, 50, 66, 70, 117–118, 122–125, 127–128, 206
light, 175–183. *See also* color; Newton, *Opticks*
modification theories of, 155
motions of, 181
particle theory of, 8–9, 135, 152–169, 175, 181–183
wave theory of, 10, 54, 135, 155–160, 166, 169, 182
Locke, John, 4–5, 28–38, 49, 52, 55–56
The Conduct of the Understanding, 32
Essay Concerning Human Understanding, 17, 28–36
on types of knowledge, 30–36
"Thoughts concerning Education," 31

McGuire, J. E., 4, 7, 13, 109–128, 140
Mach, Ernst, 69, 75
machine, world as. *See* mechanism
MacIntyre, Alasdair, 218–220, 228, 235–236
McMullin, Ernan, 139–141
Macrobius, 96
Maimonides, Moses, on emanative effect, 129n11
makom, 96, 123
Mandelbaum, Maurice, 139
Manuel, Frank, 91
Marx, Karl, 233–234
mass
and gravity, 188–202, 207
in Huygens' thought, 24–25
vs. weight, 25, 50, 188
mathematical principles, 3, 52, 168. *See also* natural philosophy
matter, 35–40, 85, 198–200, 230. *See also* gravity; inertia; inertial forces
concept of for Kant and Newton, 185–209
as object of knowledge, 188–191
measuring quantity of. *See* gravity, as essential to matter
as movable in space, 191–195, 207
relation of to absolute space and time, 77–78, 81–88

relation of to gravity. *See* gravity, as essential to matter
measure
natural, 61–64
sensible, of space, time, and motion, 6, 57–65, 73, 79–82
mechanical philosophy. *See* corpuscularian philosophy
mechanism, after Newton, 211–225, 227–237
Medieval thought. *See* Scholasticism
Mercury, 141, 143
momentum, 62, 207. *See also* conservation of momentum
Montesquieu, Baron, 12
Moore, G. E., 216, 228
morality, 211–225, 227–237. *See also* ethical relativism; ethics
moral codes, 215
moral law, 218, 233
moral obligation, 214, 220–221
objective and subjective, 217–220, 228–237
as science, 33
More, Henry, 7, 15n18, 96, 105, 111–115
motion, 19–39. *See also* laws of motion, Newton's
absolute, 10, 55, 58–75, 77–88, 190–191, 195–198, 204
circular, 6, 21–24, 58–59, 81, 83, 200, 204 (*see also* bucket, rotating; centrifugal force; centripetal force)
first law of. *See* inertia; inertial forces
in Huygens' thought, 21–26
a property of material objects, 82–83
relative, of objects, 60, 64, 79–81, 204, 209
second law of, 24, 60, 205
third law of. *See* conservation of momentum

Nagel, Ernest, 53–54
naturalistic fallacy, 214
natural language, 234–235
natural law, 211–212, 214–215, 218, 229–232, 237
natural philosophy, 3–4, 20, 231–237
functional autonomy of, 4
methodology and, 182
principles of as mathematical, 5, 39
question of as scientific, 31–33

Newton, Sir Isaac, *passim*
  Advertissement au lecteur, 96, 186
  General Scholium, 3–4, 6, 93, 97–98,
    103, 118–119, 139, 212–213
  "De gravitatione," 35–37, 51, 53–54, 96,
    103, 109, 116, 128
  laws of motion. *See* laws of motion,
    Newton's
  Opticks, 3–5, 8–9, 12–14, 36–39, 53, 54,
    70, 135–169, 175–183, 186–187, 199–
    200, 206
  *Philosophiae Naturalis Principia Mathemat-
    ica*, 1–14, 22, 24, 30–31, 37–40, 49, 53,
    55, 60, 69, 92, 96, 135–139, 143–149,
    175, 182–183, 185–186, 189–195, 198–
    200, 202, 208, 211–212
  Queries 28 and 29, of *Opticks*, 135–169,
    175
  Query 31, of *Opticks*, 175, 177, 187, 199,
    206
  "Rules of Reasoning in Philosophy"
    (from *Principia*), 3, 52, 136–137
  Scholium to the Definitions, 57–75, 77–
    82, 95, 144, 203
  "Tempus et Locus," 91–97, 100, 102,
    116, 120, 127
  Yahuda manuscript, 95, 98
Newtonian worldview, 11–14, 227–237
Newton's rings, 153, 175–176
Nicholas of Cusa, 232
Nietzsche, Friedrich
  on moral tradition, breakdown of, 219–
    220
  on creativity and values, 233

Oldenburg, Henry, 136, 148, 151, 158,
  181

passive principles, 10, 38–39, 53, 208
  See also active principles; inertia; inertial
    forces
perpetual motion, impossibility of, 21
phenomena, Newton's account of, 135–
  169, 175–183. *See also* deductions of
  propositions from phenomena
Plato, 115–116, 221, 224
Plotinus, 93–94
positivism, 54, 55
Popkin, Richard, 232,
predicates of being, 91–106, 109–128
Prichard, H. A., 217, 219, 228

primary qualities. *See* qualities of bodies,
  primary
probability, in Newton's methodology,
  135–169, 182–183
proof by experiment, question of for
  Newton, 135–169, 177–183. *See also* experimental philosophy; hypothesis
properties. *See* affections of being *qua*
  being
propositions, Newton's account of, 139
propria. *See* accidents, proper
Ptolemy, 190–191

qualities, space and time as, 101
qualities of bodies, 139–147
  primary, 28–30, 33, 49, 214
  quantifiable, 183
  secondary, 28–30, 49, 214
quantities, relative. *See* measure, sensible

Ramsey sentences, 67, 71–72
realism, about space and time, 6, 57–75,
  77–88. *See also* representationalism
reason
  and ethics, 212–214 (*see also* ethics)
  practical, 222–225, 227–237
  theoretical, 231–237
refraction, 145–146, 150, 153, 157, 162,
  166, 181
refrangibility, 27, 30, 52, 138, 141, 145–
  146, 150, 152–153, 158, 162
Reichenbach, Hans, 55–56
Reid, Thomas, 12
relationalism, 57, 61, 73, 77. *See also* representationalism
relationism. *See* relationalism
relativism, cultural, as justification for
  ethical relativism. *See* ethical relativism
relativism, ethical. *See* ethical relativism
relativity
  Galilean, 4, 21–22, 50, 203, 205, 209
  general, 65–66, 85
  special, 65–67, 77, 209
representationalism, vs. realism, 6, 57–
  75, 77–88
rods. *See* clocks, and rods
Ross, W. D., 217, 219, 228
rotation. *See* bucket, rotating; motion,
  circular
Russell, Bertrand, 216

Sabra, A. I., 153
Saturn. *See* Jupiter, and Saturn
Scholasticism, 35–37, 92, 105, 112, 124–127, 140, 211–212
Scholium, General, and to the Definitions (from Newton's *Principia*). *See* Newton, General Scholium, Scholium to the Definitions
secondary qualities, of bodies. *See* qualities of bodies, secondary
sempiternity. *See* God, sempiternity of
sensible measures of real quantities. *See* measure, sensible
Sklar, Lawrence, 6, 77–88
Snell's law, 179
Socrates, 115–116, 119, 121, 126, 224
solar time, 60, 66, 67, 69, 80
space, absolute, 6, 9–10, 57–65, 68, 77–88, 91–106, 190–191, 200–201, 203–207. *See also* infinity
space and time,
    as accidents of God, 122–124, 127
    as affections of being. *See* affections of being *qua* being, space and time as
    causation of. *See* causation, of space and time
    as caused by God. *See* God, as cause of space and time
    as conditions of being or existence, 110, 126
    as emanative effects of God. *See* emanative causation, of space and time by God
    existence of bodies in, 124–128
    as framework for primary qualities, 5–7
    as immaterial entities, 81–88
    infinity of, 91, 94–95, 100–106 (*see also* God, infinity of; infinity; space, absolute; time, absolute)
    location of bodies in, 185, 188, 191
    measuring real vs. absolute. *See* measure, sensible
    as predicates of pure existence, 91–106, 109–128
    as quantities of existence, 125–126
    as "real" beings, 110–115, 118
    relation to God in question, 117–128 (*see also* God, as cause of space and time)
    as substance or accident, 54, 91–106, 109–128

space-time, 6, 59, 61, 65–68, 71–75, 78, 84–88, 208, 223
    absolute, 77, 85
    affine structure as cause, 61, 65, 84–85
Spinoza, Baruch, 99
statistical mechanics, 62–63
Stein, Howard, 4–5, 49–56, 205–206
Stillingfleet, Locke's correspondence with, 31–33
subjectivism, 216–225, 227–232, 237
substance, 35–40, 100–106, 205, 230. *See also* extension; matter
    accident vs., 109–128 (*see also* space and time, as substance or accident)
    bodies as, 125–127
    God as possessing, 95–96, 100
    Newton's theory of, as space endowed with particular qualities, 54
substantivalism. *See* realism, about space and time
synthesis, in Newton's thought, 151–152, 177, 179

Taylor, Charles, 234–235
Thrasymachus, 221
time, absolute, 6, 10, 54–55, 57–75, 77–88, 91–106, 203. *See also* eternity; God, eternity of
Toulmin, Stephen, 68–69, 71, 75
Truesdell, Clifford, 24

unity, as coextensive with being, 110–111, 115–116, 120, 126 (*see also* Aristotle, on unity and being)
    vs. existence, 121
    of Socrates, 119, 126

values, moral. *See* morality
velocity, 23–24, 74
    absolute, 78, 80, 84, 205 (*see also* acceleration; motion)
Venus, 141, 143, 189
virtue, 211
voluntarism, 106, 227–229, 232–234, 237

wave theory of light. *See* light, wave theory of
Wolfson, Harry, 121